Student Solutions Manual for

VC Gerd

Scheaffer, Mendenhall, and Ott's

ELEMENTARY SURVEY SAMPLING

Fifth Edition

Prepared by Miseon Song

University of Florida

Duxbury Press
An Imprint of Wadsworth Publishing Company
I(T)P° An International Thomson Publishing Company

Belmont • Albany • Bonn • Boston • Cincinnati • Detroit • London • Madrid •
Melbourne
Mexico City • New York • Paris • San Francisco • Singapore • Tokyo • Toronto •
Washington

CONTENTS

CHAPTER 2
A REVIEW OF SOME BASIC CONCEPTS

2.5 $s^2 = \dfrac{1}{n-1}\sum(y_i - \bar{y})^2 = \dfrac{1}{n-1}\sum\left(y_i^2 - 2y_i\bar{y} + \bar{y}^2\right)$

$\qquad = \dfrac{1}{n-1}\left(\sum y_i^2 - 2\bar{y}\sum y_i + n\bar{y}^2\right)$

$\qquad = \dfrac{1}{n-1}\left(\sum y_i^2 - 2n\bar{y}^2 + n\bar{y}^2\right) = \dfrac{1}{n-1}\left(\sum y_i^2 - n\bar{y}^2\right)$

2.7

	Summary Statistics			
	Calories		Cost in Dollars	
	w/Hydra	w/o Hydra	w/ Hydra	w/o Hydra
Mean	64.78	64.62	.294	.27
Median	66.0	63.5	.260	.25
Stdev	8.51	9.09	.097	.05
Q_1	60	60	.230	.225
Q_3	70	70	.345	.320
Min	50	50	.220	.220
Max	80	80	.520	.350
Range	30	30	.300	.130

Scatter plot of cost vs. calories

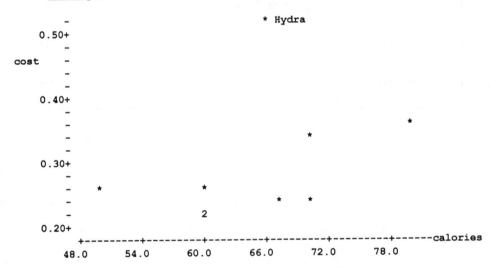

1

<u>Scatter plot of cost vs. calories (without Hydra)</u>

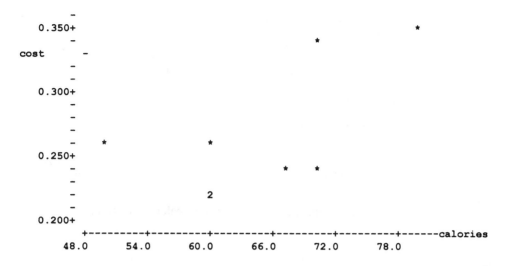

```
        -
0.350+                                              *
cost    -                              *
        -
        -
0.300+
        -
        -
        -
        -       *           *
0.250+
        -
        -                          *       *
        -                   2
0.200+
        +---------+---------+---------+---------+---------+------calories
      48.0      54.0      60.0      66.0      72.0      78.0
```

(a)   Mean is a good summary number for typical calories per serving.
Standard deviation is a good summary number for the variation in the calories.

<u>Boxplot of calories</u>

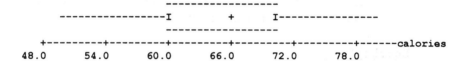

```
                      -------------------
          ----------------I      +      I----------------
                      -------------------
      +---------+---------+---------+---------+---------+------calories
    48.0      54.0      60.0      66.0      72.0      78.0
```

(b)   Since there is an extreme value, median is a good summary number for typical
cost per serving, and IQR (Q_3 - Q_1) is a good summary for the variation in
costs.

<u>Boxplot of cost</u>

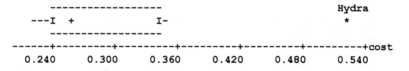

```
          -------------------                    Hydra
      ---I    +             I-                      *
          -------------------
    ------+---------+---------+---------+---------+---------+cost
      0.240     0.300     0.360     0.420     0.480     0.540
```

(c)   no, no
(d)   On the average calories per serving; not much impact
On the standard deviation: increased
On the average cost per serving: slightly decreased
On the standard deviation of the cost per serving: decreased

2

Box plot of calories (without Hydra)

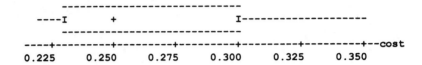

```
                              -------------------
          -----------------I        +           I----------------
                              -------------------
    +---------+---------+---------+---------+---------+------calories
   48.0      54.0      60.0      66.0      72.0      78.0
```

Box plot of cost (without Hydra)

```
                  -----------------------------
      ----I        +                   I-------------------
                  -----------------------------
    ----+---------+---------+---------+---------+---------+--cost
   0.225     0.250     0.275     0.300     0.325     0.350
```

(e) There is no influential drink on the average calories per saving.

2.9 Summary Statistics

Area	N	mean	median	stdev	Q_1	Q_3	$Q_3 - Q_1$	R=max-min
U.S	10	25.10	12.50	22.02	7.50	51.25	43.75	54.00
U.S.& Foreign	10	4.80	1.00	7.18	0.00	10.00	10.00	19.00

(a) According to stem plot, there are two modes (30's, and 50's). There is no single number to represent this data.

Box plot of U.S.

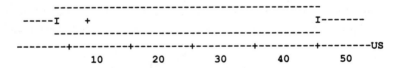

```
          ------------------------------------------------
    -----I    +                                    I-------
          ------------------------------------------------
    --------+---------+---------+---------+---------+--------US
           10        20        30        40        50
```

Stem plot of U.S.

```
Stem-and-leaf of US       N  = 10
Leaf Unit = 1.0

     3      0 368
    (3)     1 023
     4      2
     4      3 7
     3      4
     3      5 057
```

(b) Since the distribution of U.S. & Foreign is skewed to the right and single mode, median seems to be a typical value to represent this data.

Box plot of U.S & Foreign

```
Stem-and-leaf of US & For   N  = 10
Leaf Unit = 1.0

     5      0 00000
     5      0 23
     3      0
     3      0
     3      0 8
     2      1
     2      1
     2      1
     2      1 6
     1      1 9
```

Stem plot of U.S. & Foreign

```
       -----------------------
    I  +                      I--------------------------------
       -----------------------
     +---------+---------+---------+---------+---------+------US & Foreign
    0.0       3.5       7.0      10.5      14.0      17.5
```

(c) No. This is two modes data.

2.11 Parallel boxplot among players

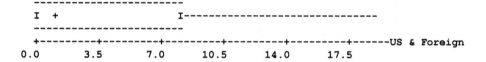

```
    Player
                                        ------------
    Ruth                  --------------I    +    I-------
                                        ------------
                          -------------------
    Gehrig------------------I         +        I-------
                          -------------------
                          ---------------
    Mantle          ------I     +     I--------------
                          ---------------
                      ---------------
    Maris       --------I    +     I---------------------
                      ---------------
              --+---------+---------+---------+---------+---------+----HomeRuns
                0        12        24        36        48        60
```

4

Player	N	mean	median	stdev	Q_1	Q_3	$Q_3 - Q_1$	R=max-min
Ruth	15	43.93	46.00	11.25	35.00	54.00	19.00	46.00
Gehrig	17	29.00	32.00	16.11	18.00	34.50	25.50	49.00
Mantle	18	29.78	28.50	11.94	20.50	37.75	17.25	28.50
Maris	7	29.00	26.00	17.71	13.00	39.00	26.00	53.00

We select Ruth as the greatest player since Ruth's home run distribution shifted to the right (see above box plot).

Also, the table above indicates Ruth was the most consistent player as he had little standard deviation and with Maris was the least consistent player he had the most standard deviation, IQR and R.

2.13 **(a)**

State	R	M	Ave
Alaska	471	4	8492.6
New York	9771	108	11053.1
Rhode Island	628	7	11146.5
Florida	9980	114	11422.8
California	22253	258	11593.9
Total	43103	491	

(R : thousands, M: billions)

(b) $\bar{x} = 491,000,000 / 43103 = 11391.3$

(c) yes, $\bar{x}$ is close to 10,000.

2.15 Stemplot and Boxplot indicate that distribution of SAT is almost symmetric and the distribution of Percent is skewed to the right. And scatter plot shows that SAT and Percent are negatively correlated.

Summary statistics of SAT and Percent

	N	MEAN	MEDIAN	STDEV	MIN	MAX	Q_1	Q_3
SAT	51	944.10	926.00	66.93	832	1093	886	997
Percent	51	35.76	26.00	26.19	4	81	11	61

Box plot

```
MTB > boxplot 'SAT'
```

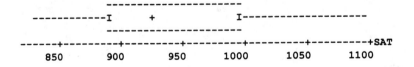

```
MTB > boxplot 'Percent'

            ---------------------------------
       ----I           +                I--------------
            ---------------------------------
       +---------+---------+---------+---------+---------+------Percent
            0         15        30        45        60        75
```

Stemplot by Minitab (SAT and Percent)

```
Stem-and-leaf of SAT    N = 51
Leaf Unit = 10

     1      8 3
     4      8 444
     8      8 6777
    19      8 88888999999
    22      9 011
    (5)     9 22223
    24      9 45
    22      9 6
    21      9 888999999
    12     10 0001
     8     10 22233
     3     10 4
     2     10 7
     1     10 9


Stem-and-leaf of Percent   N  = 51
Leaf Unit = 1.0
     1      0 4
     9      0 55566899
    19      1 0011122223
    22      1 678
    24      2 22
    (3)     2 569
    24      3
    24      3
    24      4 14
    22      4 789
    19      5 4
    18      5 5778
    14      6 01244
     9      6 778
     6      7 14
     4      7 559
     1      8 1
```

6

Scatter plot between SAT and Percent

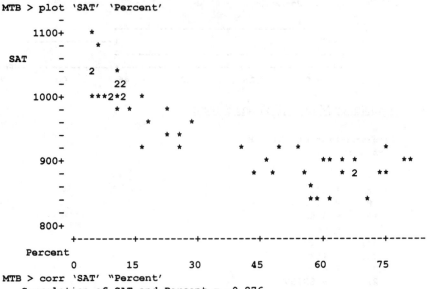

```
MTB > plot 'SAT' 'Percent'
        -
  1100+    *
        -       *
 SAT    -
        -    2   *
        -       22
  1000+    ***2*2   *
        -      *  *
        -         *        *
        -            * *
        -             * *
        -         *      *       *     * *         *
   900+                        *      ** * *       **
        -                        *  *    *     * 2   **
        -                                   *
        -                                 ** *    *
        -
   800+
        +---------+---------+---------+---------+---------+------
   Percent
        0        15        30        45        60        75
MTB > corr 'SAT' "Percent'
   Correlation of SAT and Percent = -0.876
```

2.17

	Aspirin	Placebo
Exercise Vigorously		
Yes	7 910	7861
No	2997	3060
Total	10907	10921

	Aspirin	Placebo
Cigarette smoking		
Never	5431	5488
Past	4373	4301
Current	1213	1225
Total	11017	11014

(a) Compare column totals in each table.
Since 10907 and 11092 are nearly same and 11017 and 11014 are nearly same, randomization scheme did a good job in controlling these variables.

(b) no. $\dfrac{5431}{11017} = .49$, $\dfrac{5488}{11014} = .50$ are nearly same

(c) no. $\dfrac{2997}{10907} = .27$, $\dfrac{3060}{10921} = .28$ are nearly same

7

2.19

Stroke	Aspirin	Placebo
Yes	119	98
No	10881	10902
Total	11000	11000

No, $\dfrac{119}{11000} = .0108$ $\dfrac{98}{11000} = .0089$ are close.

Comparing two ratios $\dfrac{.021727}{.012636} = 1.72,$ $\dfrac{.0089}{.0108} = .82$, we can find Aspirin is more effective as a possible prevention for heart attacks than for strokes.

2.21 $p(u_1) = p(u_2) = \cdots = p(u_N) = 1/N$

$$\sigma^2 = V(y) = E(y - \mu)^2 = \sum_y (y - \mu)^2 p(y) = \frac{1}{N} \sum_{i=1}^{N} (u_i - \mu)^2$$

CHAPTER 3
ELEMENTS OF THE SAMPLING PROBLEM

3.23 **(a)** 1 rating = 92.1 million × 0.01 = 921,000 households

(b) 1 share is going to be larger than 1 rating
1 rating = 1% of sampled households (4,000) that have a TV set and turned to the program
1 share = 1% of viewing households that have a TV set and turned to the program where a viewing household is a household that has at least one TV set turned on.

(c) 92.1 million × .217 = 19.98 million households.

3.25 **(a)**

Target Population	51%	12%	9%
High risk cities	57.9%	33.8%	20.7%
National	58.4%	13.5%	8.3%

The % of the High risk cities of female, black, and Hispanic is higher than the % of the target population. The randomized survey did not well in fairly representing those three groups.

But, in national survey, only the % of female group is much higher than that of target population. Thus, randomization did a good job for black and Hispanic group in national survey.

(b) High risk cities are not the typical cities of the population.

3.29

	NS	EX	FS	CS	Total
A lot	7213	2693	75	857	10838
Somewhat	2482	1861	109	1102	5554
A Little	744	542	27	298	1611
Don't care	1878	1550	119	1312	4859
Total	12317	6646	330	3569	22862
% A lot	.59	.41	.23	.24	

Care about staying away from marijuana

(a) 7213 / 12317 = .59

(b) 857 / 3569 = .24

(c) 7213 / 10838 = .67

(d) 1878 / 4859 =.39

(e) Non smokers care more about staying away from marijuana than current smokers (59%, 24%, respectively). Also from (a), and (b), among those who care a lot about staying away from marijuana, 59% was non smokers, while 24% was current smokers.

CHAPTER 4

SIMPLE RANDOM SAMPLING

4. 1

Possible samples	{0,1}	{0,2}	{0,3}	{0,4}	{1,2}	{1,3}	{1,4}	{2,3}	{2,4}	{3,4}
sample mean $\bar{y}$	0.5	1.0	1.5	2.0	1.5	2.0	2.5	2.5	3.0	3.5

where $\bar{y} = \frac{1}{n}\sum y_i$

The probability distribution of sample mean $\bar{y}$ is

$\bar{y}$	0.5	1.0	1.5	2.0	2.5	3.0	3.5
$p(\bar{y})$	0.1	0.1	0.2	0.2	0.2	0.1	0.1

$$E(\bar{y}) = \sum_{\bar{y}} \bar{y}p(\bar{y})$$
$$= (0.5)(.1) + (1.0)(.1) + \cdots + (3.0)(.1) + (3.5)(.1) = 2$$

$$E(\bar{y}^2) = \sum_{\bar{y}} \bar{y}^2 p(\bar{y})$$
$$= (0.5)^2(.1) + (1.0)^2(.1) + \cdots + (3.0)^2(.1) + (3..5)^2(.1) = 4.75$$

$$V(\bar{y}) = E(\bar{y}^2) - E(\bar{y})^2 = 4.75 - 4 = .75$$

The probability distribution of y is

y	0	1	2	3	4
$p(y)$	.2	.2	.2	.2	.2

$$E(y) = \sum_{y} yp(y) = 0(.2) + 1(.2) + 2(.2) + 3(.2) + 4(.2) = 2$$

$$E(y^2) = \sum_{y} y^2 p(y) = 0^2(.2) + 1^2(.2) + 2^2(.2) + 3^2(.2) + 4^2(.2) = 6$$

$$\sigma^2 = V(y) = E(y^2) - E(y)^2 = 6 - 4 = 2$$

So, $V(\bar{y}) = \frac{N-n}{N-1}\left(\frac{\sigma^2}{n}\right) = \frac{5-2}{5-1}\left(\frac{2}{2}\right) = \frac{3}{4} = .75$

4.5 $\hat{p} = \dfrac{\sum y_i}{n} = \dfrac{25}{30} = \dfrac{5}{6} = 0.83$

$B = 2\sqrt{\dfrac{\hat{p}\hat{q}}{n-1}\left(\dfrac{N-n}{N}\right)} = 2\sqrt{\dfrac{(5/6)(1/6)}{29}\left(\dfrac{300-30}{300}\right)} = .131$

4.7 $\hat{\mu} = \bar{y} = 12.5$

$B = 2\sqrt{\dfrac{s^2}{n}\left(\dfrac{N-n}{N}\right)} = 2\sqrt{\dfrac{1252}{100}\left(\dfrac{10000-100}{10000}\right)} = 7.04$

4.9 $N = 10{,}000 \quad n = 500$

$\hat{\mu}_1 = \bar{y}_1 = 2.3$

$\hat{\mu}_2 = \bar{y}_2 = 4.52$

$B_1 = 2\sqrt{\dfrac{s_1^2}{n_1}\left(\dfrac{N_1-n_1}{N_1}\right)} = 2\sqrt{\dfrac{.65}{500}\left(\dfrac{10000-500}{10000}\right)} = .070$

$B_2 = 2\sqrt{\dfrac{s_2^2}{n_2}\left(\dfrac{N_2-n_2}{N_2}\right)} = 2\sqrt{\dfrac{.97}{500}\left(\dfrac{10000-500}{10000}\right)} = .086$

4.11 $N = 1000,\, n = 10$

$\bar{y} = \dfrac{\sum y_i}{n} = \dfrac{20}{10} = 2.0$

$s^2 = \dfrac{\sum(y_i - \bar{y})^2}{n-1} = \dfrac{\sum y_i^2 - n\bar{y}^2}{n-1} = \dfrac{60-10(4)}{9} = \dfrac{20}{9} = 2.22$

$\hat{\mu} = \bar{y} = 2$

$B = 2\sqrt{\dfrac{s^2}{n}\left(\dfrac{N-n}{N}\right)} = 2\sqrt{\dfrac{2.22}{10}\left(\dfrac{1000-10}{1000}\right)} = .938$

4.13 $B = .02, \quad D = B^2 / 4 = (.02)^2 / 4 = .0001$

$n = \dfrac{Npq}{(N-1)D + pq}$

$= \dfrac{99000\,(.43)\,(.57)}{98999\,(.0001) + .43\,(.57)} = 2391.8 \approx 2392$

4.15 $\bar{y} = 2.1 \quad s = .4 \quad N = 200, \quad n = 20$

$\hat{\mu} = \bar{y} = 2.1$

$$B = 2\sqrt{\frac{s^2}{n}\left(\frac{N-n}{N}\right)} = 2\sqrt{\frac{(.4)^2}{20}\left(\frac{200-20}{200}\right)} = .17$$

4.17 $\hat{\mu} = \frac{1}{n}\sum y_i = \frac{1}{8}40.1 = 5.01, \quad s^2 = 1.65$

$$B = 2\sqrt{\frac{s^2}{n}\left(\frac{N-n}{N}\right)} = 2\sqrt{\frac{1.65}{8}\left(\frac{98-8}{98}\right)} = .871$$

4.19 $B = .08, \quad D = B^2/4 = (.08)^2/4 = .0016$

$$n = \frac{N pq}{(N-1)D + pq} = \frac{621(.2)(.8)}{620 \,(.0016) + .2(.8)} = 86.25 \approx 87$$

4.21 By using s^2 to estimate σ^2 in Equation (4.14),

$$n = \frac{N s^2}{(N-1)D + s^2} = \frac{1500 \,(136)}{1499(4) + 136} = 399.4 \approx 400$$

where

$$D = \frac{B^2}{4N^2} = \frac{(1500)^2}{4(1500)^2} = 4$$

4.23 $n = 811, \quad \hat{p} = .57$

(a) $B = 2\sqrt{\frac{\hat{p}\hat{q}}{n-1}} = 2\sqrt{\frac{.57(.43)}{810}} = .035$

(b) $\hat{p} \pm B = .57 \pm .035$ or .535 to .605

Thus we can be confident that most Floridians disapprove of legalizing casinos in the state since interval does not include 0.5.

4.27

	# of bats	# of hits	batting average	error bound	
	N	n	$\hat{p}$	B	95% C.I.
Regular season	9864	2584	.262	.015	(.247, .277)
League Champ.	163	37	.227	.123	(.154, .350)
World Series	98	35	.357	.132	(.225, .489)

where $B = 2\sqrt{\frac{\hat{p}\hat{q}}{n-1}\left(\frac{N-n}{N}\right)}$, C.I. $= \hat{p} \pm B$

Since intervals are overlapping each other, Reggie Jackson's nickname is not justified.

4.29 $N = 1000$, $n = 15$, $\sum y_i = 260$, $s^2 = 75.2381$

$\hat{\tau} = N\bar{y} = 1000(260)/15 = \$17,333.33$

$$B = 2\sqrt{N^2\left(\frac{s^2}{n}\right)} = 2\sqrt{(1000)^2\,\frac{75.2381}{15}} = \$4479.23$$

(Ignoring the fpc)

4.31 $\hat{p}$ = proportion of the firm's account that fail to comply with stated procedures

$$\hat{p} = \frac{\sum y_i}{n} = \frac{6}{20} = .3$$

$$B = 2\sqrt{\frac{\hat{p}\hat{q}}{n-1}\left(\frac{N-n}{N}\right)} = 2\sqrt{\frac{.3(.7)}{19}\left(\frac{500-20}{500}\right)} = .206$$

An estimates confidence interval for the proportion of accounts that comply with stated procedures is $(.7 - .206, .7 + .206)$, or $.494$ to $.906$. Thus we cannot think the proportion of accounts that comply with stated procedures exceeds 80 %.

4.33

	all data		without LA		without Houston & LA	
year	mean	stdev	mean	stdev	mean	stdev
1981	19.67	37.75	10.21	9.57	8.46	7.25
1985	17.87	38.47	8.21	9.41	6.54	7.30
1990	13.60	29.78	6.29	9.53	4.08	4.94

(a) $\bar{y}_1 - \bar{y}_2 = 19.67 - 17.87 = 1.8$

$$B = 2\sqrt{\frac{s_1^2}{n_1} + \frac{s_2^2}{n_2}} = 2\sqrt{\frac{(37.75)^2}{15} + \frac{(38.47)^2}{15}} = 27.83$$

(b) $\bar{y}_2 - \bar{y}_3 = 17.87 - 13.60 = 4.27$

$$B = 2\sqrt{\frac{s_2^2}{n_2} + \frac{s_3^2}{n_3}} = 2\sqrt{\frac{(38.47)^2}{15} + \frac{(29.78)^2}{15}} = 25.12$$

(c) $\bar{y}_1 - \bar{y}_2 = 10.21 - 8.21 = 2.0$

$$B = 2\sqrt{\frac{s_1^2}{n_1} + \frac{s_2^2}{n_2}} = 2\sqrt{\frac{(9.57)^2}{14} + \frac{(9.41)^2}{14}} = 7.17$$

$$\bar{y}_2 - \bar{y}_3 = 8.21 - 6.29 = 1.92$$

$$B = 2\sqrt{\frac{s_2^2}{n_2} + \frac{s_3^2}{n_3}} = 2\sqrt{\frac{(9.41)^2}{14} + \frac{(9.53)^2}{14}} = 7.16$$

(d) $\quad \bar{y}_1 - \bar{y}_2 = 8.46 - 6.54 = 1.92$

$$B = 2\sqrt{\frac{s_1^2}{n_1} + \frac{s_2^2}{n_2}} = 2\sqrt{\frac{(7.25)^2}{13} + \frac{(7.30)^2}{13}} = 5.71$$

$$\bar{y}_2 - \bar{y}_3 = 6.54 - 4.08 = 2.46$$

$$B = 2\sqrt{\frac{s_2^2}{n_2} + \frac{s_3^2}{n_3}} = 2\sqrt{\frac{(7.30)^2}{13} + \frac{(4.94)^2}{13}} = 4.89$$

4.35 **(a)** Comparisons of the proportions who want to vote for Republican candidate for this time period involves a difference of independent proportions. The estimate and bound on the error is

$$.40-.34 \pm 2\sqrt{\frac{(.40)(.60)}{800} + \frac{(.34)(.66)}{800}}$$
$$.06 \pm .048 \text{ or } .012 \text{ to } .108$$

Since interval does not contain zero, there is statistical evidence that Republicans had made a significant gain in this time period.

(b) The proportion of those who want to vote for Republican candidate is dependent on the proportion of those who want to vote for Democratic candidate. The estimate and bound on the error is

$$.40-.38 \pm 2\sqrt{\frac{(.40)(.60)}{800} + \frac{(.38)(.62)}{800} + 2\frac{(.40)(.38)}{800}}$$
$$.02 \pm .062 \text{ or } -.042 \text{ to } .082$$

Since interval contains zero, one cannot say that Republican votes would outnumber Democratic votes as of September 1, 1994.

4.37 $.39-.37 \pm 2\sqrt{\frac{(.39)(.61)}{773} + \frac{(.37)(.63)}{773} + 2\frac{(.39)(.37)}{773}}$

$.02 \pm .063 \text{ or } -.043 \text{ to } .083$

Yes, the Florida poll was correct in its interpretation of the data that the race between Clinton and Bush was "too close to call".

4.39 **(a)** $\hat{p} = .22$

$$B = 2\sqrt{\frac{\hat{p}\hat{q}}{n-1}\left(\frac{N-n}{N}\right)} = 2\sqrt{\frac{(.22)(.78)}{81}\left(\frac{1400-82}{1400}\right)} = .0893$$

(b) $\hat{p} = .18+.19+.04+.22 = .63$

$$B = 2\sqrt{\frac{\hat{p}\hat{q}}{n-1}\left(\frac{N-n}{N}\right)} = 2\sqrt{\frac{(.63)(.37)}{81}\left(\frac{1400-82}{1400}\right)} = .1041$$

(c) $\hat{p} = .10$

$$B = 2\sqrt{\frac{\hat{p}\hat{q}}{n-1}\left(\frac{N-n}{N}\right)} = 2\sqrt{\frac{(.1)(.9)}{45}\left(\frac{1400-45}{1400}\right)} = .0880$$

(d) $\hat{p} = .35+.05+.35+.15 = .90$

$$B = 2\sqrt{\frac{\hat{p}\hat{q}}{n-1}\left(\frac{N-n}{N}\right)} = 2\sqrt{\frac{(.9)(.1)}{45}\left(\frac{1400-45}{1400}\right)} = .0880$$

4.41 $n = 64, \hat{\mu} = \$18,300, s = 400$

$$B = 2\sqrt{\frac{s^2}{n}} = 2\sqrt{\frac{400^2}{64}} = \$100$$

(18300-100,18300+100) = (18200, 18400) is approximately a 95% C.I. for the population mean.

We can support the claim that secretaries are being paid low wages. Needed assumption: sample mean has a normal distribution.

4.43 **(a)** $\hat{\mu} = 19.9, B = 2\sqrt{\frac{s^2}{n}} = 2\sqrt{\frac{(17.3)^2}{2001}} = .773$

95 % confidence interval:
$\hat{\mu} \pm B = (19.9 - .773, 19.9 + .773) = (19.127, 20.673)$

(b) $\hat{\mu} = 18.0, B = 2\sqrt{\frac{s^2}{n}} = 2\sqrt{\frac{(15.3)^2}{1822}} = .717$

95 % confidence interval:
$\hat{\mu} \pm B = (18.0 - .717, 18.0 + .717) = (17.283, 18.717)$

Since C.I. estimates do not overlap, true means probably differ.

(c) $\quad B = 2\sqrt{\dfrac{s^2}{n}}, \quad \text{C.I.} \; = \; (\hat{\mu} \pm B)$

	Large Firms		Small Firms	
	pre	post	pre	post
n	2001	2001	1822	1822
μ	5.3	12.7	4.9	11.2
s	3.6	15.5	3.7	13.5
B	.161	.693	.173	.633
C.I.	5.319	12.007	4.727	10.567
	5.461	13.393	5.073	11.833

(d) $\quad \hat{p} = \dfrac{307 + 1930}{307 + 1930 + 901 + 426 + 259} = \dfrac{2237}{3823} = .585$

$$B = 2\sqrt{\hat{V}(\hat{p})} = 2\sqrt{\dfrac{\hat{p}\hat{q}}{n-1}} = 2\sqrt{\dfrac{.585(.415)}{3822}} = .016$$

CHAPTER 5
STRATIFIED RANDOM SAMPLING

5.1

	Stratum			
	I	II	III	IV
N_i	65	42	93	25
n_i	14	9	21	6
# of acct.	4	2	8	1
$\hat{p}_i$	.286	.222	.381	.167

$N = 225$

The estimate of the proportion of delinquent accounts is , using Equation (5.13),

$$\hat{p}_{st} = \frac{1}{N}\sum N_i\hat{p}_i = \frac{1}{225}\left[65(.286) + 42(.222) + 93(.381) + 25(.167)\right] = .30$$

The estaiamted variance of $\hat{p}_{st}$ is, by Equation (5.14),

$$\hat{V}(\hat{p}_{st}) = \frac{1}{N^2}\sum N_i^2\left(\frac{N_i - n_i}{N_i}\right)\left(\frac{\hat{p}_i\hat{q}_i}{n_i - 1}\right) = .0034397$$

with a bound on the error of estimation

$$B = 2\sqrt{\hat{V}(\hat{p}_{st})} = .117$$

5.3

	Stratum		
	I	II	III
N_i	132	92	27
n_i	18	10	2
$\bar{y}_i$	8.83	6.7	4.5
s_i^2	81.56	50.46	24.50

$N = 251$

The estiamte of the total number of man-hours lost during the given month is, from Equation (5.3),

$$\hat{\tau} = N\bar{y}_{st} = 132(8.83) + 92(6.7) + 27(4.5) = 1903.9$$

The estimated variance of $\hat{\tau}$ is, from Equation (5.4),

$$V(\hat{\tau}) = \hat{V}(N\bar{y}_{st}) = \sum N_i^2\left(\frac{N_i - n_i}{N_i}\right)\frac{s_i^2}{n_i} = 114519.9$$

with a bound on the error of estimation

$$B = 2\sqrt{\hat{V}(\bar{y}_{st})} = 676.8$$

(c) There are two outliers, Phoenix and Atlanta in MSA.

5.5

	Stratum			
	I	II	III	
N_i	112	68	39	$N = 219$
c_i	9	25	36	
σ_i^2	2.25	3.24	3.24	

$$n = \frac{\left(\sum N_k \sigma_k / \sqrt{c_k}\right)\left(\sum N_i \sigma_i \sqrt{c_i}\right)}{N^2 D + \sum N_i \sigma_i^2}$$

$$B = 2\sqrt{V(\bar{y}_{st})} \text{ or } V(\bar{y}_{st}) = \frac{B^2}{4} = D = 0.1$$

$$\sum\left(\frac{N_k \sigma_k}{\sqrt{c_k}}\right) = \frac{112(1.5)}{3} + \frac{68(1.8)}{5} + \frac{39(1.8)}{6} = 92.18$$

$$\sum N_i \sigma_i \sqrt{c_i} = 112(1.5)(3) + 68(1.8)(5) + 39(1.8)(6) = 1537.2$$

$$\sum N_i \sigma_i^2 = 112(2.25) + 68(3.24) + 39(3.24) = 598.68$$

$$n = \frac{92.18(1537.2)}{219^2(.1) + 598.68} = 26.3 \approx 27$$

To allocate the $n = 27$ to the three strata, use

$$n_i = n\frac{N_i \sigma_i / \sqrt{c_i}}{\sum N_k \sigma_k / \sqrt{c_k}}$$

Then

$$n_1 = 27\frac{112(1.5)/3}{92.18} = 16.40$$

$$n_2 = 27\frac{68(1.8)/5}{92.18} = 7.17$$

$$n_3 = 27\frac{39(1.8)/6}{92.18} = 3.43$$

Rounding off yields: $n_1 = 16$, $n_2 = 7$, $n_3 = 3$
This adds to total sample size of 26, not 27. Add 1 to one of the sample sizes to

achieve n = 27. Add to stratum 3 because 3.43 is closer to the next higher integer than any other sample sizes.

5.7 $n_i = n \dfrac{N_i \sigma_i}{\sum N_i \sigma_i}$

s_i will be used to estimate σ_i.

$$N_1 \sigma_1 = 55\sqrt{105.14} = 563.96$$
$$N_2 \sigma_2 = 80\sqrt{158.20} = 1006.22$$
$$N_3 \sigma_3 = 65\sqrt{186.13} = 886.79$$

$$n_1 = 50\frac{563.96}{2456.97} = 11.48, \quad n_2 = 20.48, \quad n_3 = 18.05$$

Rounding off yields: $n_1 = 11, \quad n_2 = 20, \quad n_3 = 18$
which gives a total sample size of 49, not 50. Add 1 to n_1 because n_1 is closer to the next higher integer than the other two sample sizes. Thus, we allocate $n_1 = 12$, $n_2 = 20, \quad n_3 = 18$

5.9 Using Equation (5.10),

$$n = \frac{\left(\sum N_i \sigma_i\right)^2}{N^2 D + \sum N_i \sigma_i^2} = \frac{(2456.97)^2}{200(4) + 30537.04} = 31.68 \approx 32$$

where
$$D = B^2 / 4 = 16 / 4 = 4$$
Use s_i^2 to estimate σ_i^2.
$$\sum N_i s_i = 55\sqrt{105.14} + 80\sqrt{158.20} + 65\sqrt{186.13} = 2456.97$$

5.11 $n = \dfrac{\left(\sum N_i \sigma_i\right)^2}{N^2 D + \sum N_i \sigma_i^2}$

$$N^2 D = \frac{B^2}{4} = \frac{(5000)^2}{4} = (2500)^2$$

Use s_i^2 to estimate σ_i^2.
$$\sum N_i s_i = 86\sqrt{1071.79} + 72\sqrt{16974.28} + 30\sqrt{72376.30} = 24476.20$$

$$\sum N_i s_i^2 = 86(1071.79) + 72(9054.18) + 52(16794.28) + 30(72376.30)$$
$$= 3788666.11$$

$$n = \frac{24476.20^2}{(2500)^2 + 3788666.11} = 59.68 \approx 60$$

5.13

	Stratum				
	I	II	III	IV	Tot
N_i	97	43	145	68	353
p_i	.9	.9	.5	.5	
c_i	4	4	8	8	
$N_i \sqrt{p_i q_i / c_i}$	14.55	6.45	25.63	12.02	58.65
w_i	.248	.110	.437	.205	
$N_i^2 p_i q_i / w_i$	3413.63	1513.26	12027.53	5640.50	22594.92
$N_i p_i q_i$	8.73	3.87	36.25	17.00	65.85
n_i	39	17	69	33	158

where

$$w_i = \frac{N_i \sqrt{p_i q_i / c_i}}{\sum N_k \sqrt{p_k q_k / c_k}}$$

$$D = \frac{B^2}{4} = \frac{.05^2}{4} = (.025)^2$$

$$n = \frac{\sum N_i^2 p_i q_i / w_i}{N^2 D + \sum N_i p_i q_i} = \frac{22594.92}{(353)^2 (.025)^2 + 65.85} = 157.20 \approx 158$$

$$n_i = n w_i$$

5.15   Total cost $= \sum n_i c_i = 400$

where c_i is the cost of obtaining one observation from stratum i

But, $c_i = c_2 = \dfrac{c_3}{2} = \dfrac{c_4}{2}$

Writing the equation in terms of c_1 only gives

$$n_1 c_1 + n_2 c_2 + n_3 c_3 + n_4 c_4 = 400$$
$$n_1 c_1 + n_2 c_1 + 2 n_3 c_1 + 2 n_4 c_1 = 400$$
$$c_1 (n_1 + n_2 + 2 n_3 + 2 n_4) = 400$$
$$n_1 + n_2 + 2 n_3 + 2 n_4 = 400 / c_1 = 100$$

$$nw_1 + nw_2 + 2nw_3 + 2nw_4 = 100$$
$$n(w_1 + w_2 + 2w_3 + 2w_4) = 100$$

Using the sampling fractions from Exercise 5.13,

$$n = 100 / [.248 + .110 + 2(.437) + 2(.205)] = 60.90 \cong 61$$

$$n_1 = nw_1 = 61(.248) = 15.13 \approx 15$$
$$n_2 = nw_2 = 61(.110) = 6.71 \approx 7$$
$$n_3 = nw_3 = 61(.437) = 26.66 \approx 27$$
$$n_4 = nw_4 = 61(.205) = 12.50 \approx 12$$

Total cost = 15(4) + 4(4) + 27(8) + 12(8) = \$400.

5.17

No. of Employees	Frequency	$\sqrt{\text{Frequency}}$	Cumulative $\sqrt{\text{Frequency}}$
0-10	2	1.41	1.41
11-20	4	2.00	3.41
21-30	6	2.45	5.86
31-40	6	2.45	8.31
41-50	5	2.24	10.55
51-60	8	2.83	13.38
61-70	10	3.16	16.54
71-80	14	3.74	20.28
81-90	19	4.36	24.64
91-100	13	3.61	28.25
101-110	3	1.73	29.98
111-120	7	2.65	32.62

L = 4 strata, 32.62 / 4 = 8.155

Stratum boundaries should be as close as possible to: 8.155, 16.312, 24.468 Choose boundaries of 8.31, 16.54, 24.64.

Stratum 1: 0-40   employees
Stratum 2: 41-70 employees
Stratum 3: 71-90 employees
Stratum 4: 91-120 employees

5.19

	Stratum	
	I	II
$\bar{y}_i$	63.47	64.30
s_i^2	1.07	1.30
n_i	8	7

N_1 and N_2 are unknown. Assume that they are equal. Let N' represent both of these terms.

$$\hat{\mu} = \bar{y}_{st} = \frac{1}{N} \sum N_i \bar{y}_i = \frac{1}{2N'} \left[N\bar{y}_1 + N\bar{y}_2 \right] = \frac{1}{2}(\bar{y}_1 + \bar{y}_2) = \frac{1}{2}(63.47 + 64.30) = 63.88$$

$$B = 2\sqrt{\frac{1}{N^2} \sum N_i^2 \frac{s_i^2}{n_i}} = 2\sqrt{\frac{N'^2}{(2N')^2} \sum \frac{s_i^2}{n_i}} = 2\sqrt{\frac{1}{4} \sum \frac{s_i^2}{n_i}} = 2\sqrt{\frac{1}{4}\left(\frac{1.07}{6} + \frac{1.30}{6}\right)} = .628$$

5.21 (a) $\hat{p} = \frac{\sum y_i}{n} = \frac{6+10}{100} = .16$

$$B = 2\sqrt{\frac{\hat{p}\hat{q}}{n-1}} = 2\sqrt{\frac{.16(.84)}{99}} = .074 \quad \text{(ignoring fpc)}$$

(b) $\hat{p}_{st} = \frac{1}{N} \sum N_i \hat{p}_i = \sum \frac{N_i}{N} \hat{p}_i = .6\frac{6}{38} + .4\frac{10}{62} = .16$

$$B = 2\sqrt{\frac{1}{N^2} \sum N_i^2 \frac{\hat{p}_i \hat{q}_i}{n_i - 1}} = 2\sqrt{\sum \left(\frac{N_i}{N}\right)^2 \frac{\hat{p}_i \hat{q}_i}{n_i - 1}}$$

$$= 2\sqrt{(.6)^2\left(\frac{6}{38}\right)\left(\frac{32}{38}\right)\left(\frac{1}{37}\right) + (.4)^2\left(\frac{10}{62}\right)\left(\frac{52}{62}\right)\left(\frac{1}{61}\right)} = .081$$

5.25 The simplest cost function is of the form

$$\text{cost} = C = c_0 + \sum c_h n_h$$

Then the variance of the estimated mean $\bar{y}_{st}$ is a minimum when n_h is proportional to $N_h S_h / \sqrt{c_h}$ (Cochran, Sampling Techniques, 1963, p95, Theorem 5.6), .i.e.

$$\frac{n_h}{n} = \frac{N_h S_h / c_h}{\sum N_h S_h / c_h}$$

If cost is fixed, substitute the optimum values of in the above cost function and solve for n.

5.27 **(a)**

	Stratum		
	I	II	Total
N_i	20	26	
σ_i	25	47.5	
$N_i\sigma_i$	500	1235	1735
w_i	.29	.71	
$N_i\sigma_i^2$	12500	58662.5	71162.5

where

Use $\dfrac{\text{range}}{4}$ to estimate σ_i.

$$\sigma_1 \approx \frac{100-0}{4} = 25 \quad \text{for small plants (stratum I)}$$

$$\sigma_2 \approx \frac{200-10}{4} = 47.5 \quad \text{for large plants (stratum II)}$$

$$w_i = \frac{N_i\sigma_i}{\sum N_i\sigma_i}$$

$$w_1 = \frac{500}{1735} = .29 \qquad w_2 = \frac{1235}{1735} = .71$$

(b) $B = 100$

$$N^2 D = \frac{B^2}{4} = \frac{(100)^2}{4} = 2500$$

$$n = \frac{\left(\sum N_i\sigma_i\right)^2}{N^2 D + \sum N_i\sigma_i^2} = \frac{(1735)^2}{2500 + 71162.5} = 40.87 \approx 41$$

$$n_1 = nw_1 = 41(.29) = 11.9 \approx 12$$

$$n_2 = nw_2 = 41(.71) = 29.1 \approx 29$$

Since there is only 26 "large" plants, we allocate $n_1 = 15$, $n_2 = 26$.

5.29

	Stratum		
	I	II	Total
$\hat{p}_i$	.75	.40	
n_i	80	20	$n' = 1000$
w_i'	.8	.2	
$\left(w_i'^2 - \dfrac{w_i'}{n'}\right)\dfrac{\hat{p}_i\hat{q}_i}{n_i - 1}$	.001517	.000503	.00202
$\dfrac{w_i'(\hat{p}_i\hat{p}_{st})^2}{n'}$	.00000392	.00001568	.0000196

$$\hat{p}'_{st} = \sum w'_i \hat{p}_i = .8(.75) + .2(.40) = .68$$

$$\hat{V}(\hat{p}'_{st}) = \frac{n'}{n'-1} \sum_{i=1}^{2} \left[\left(w_i'^2 - \frac{w_i'}{n'} \right) \frac{\hat{p}_i \hat{q}_i}{n_i - 1} + \frac{w_i'(\hat{p}_i - \hat{p}'_{st})^2}{n'} \right]$$

$$= \frac{1000}{999}(.00202 + .0000196) = .00204$$

5.31

	Stratum		
	I	II	III
w_i	.5	.1	.4
No	417	29	240
	31.4	17.6	21.8
Yes	913	136	860
	68.6	82.4	78.2
Tot	1330	165	1100
	100%	100%	100%

(a) $\quad \hat{p}_{st} = \sum \frac{N_i}{N} \hat{p}_i = \sum w_i \hat{p}_i = (.5)(.686) + (.1)(.824) + (.4)(.782) = .738$

$$\hat{V}(\hat{p}_{st}) = \sum w_i^2 \frac{\hat{p}_i \hat{q}_i}{n_i - 1} \quad \text{(ignoring the fpc)}$$

$$= (.5)^2 \frac{(.686)(.314)}{1329} + (.1)^2 \frac{(.824)(.176)}{164} + (.4)^2 \frac{(.782)(.218)}{1099} = .74 \times 10^{-4}$$

(b) $\quad \hat{p}_1 - \hat{p}_2 = = .686 - .824 = -.138$

$$B = 2\sqrt{\hat{V}(\hat{p}_1 - \hat{p}_2)} = 2\sqrt{\frac{\hat{p}_1 \hat{q}_1}{n_1} + \frac{\hat{p}_2 \hat{q}_2}{n_2}} = 2\sqrt{\frac{(.686)(.314)}{1330} + \frac{(.824)(.176)}{165}} = .0645$$

(c) $\quad \hat{p}_1 - \hat{p}_2 = = .686 - .782 = -.096$

$$B = 2\sqrt{\hat{V}(\hat{p}_1 - \hat{p}_2)} = 2\sqrt{\frac{\hat{p}_1 \hat{q}_1}{n_1} + \frac{\hat{p}_2 \hat{q}_2}{n_2}} = 2\sqrt{\frac{(.686)(.314)}{1330} + \frac{(.782)(.218)}{1100}} = .036$$

CHAPTER 6

RATIO, REGRESSION, AND DIFFERENCE ESTIMATION

6.1   A scatter plot of the data shows evidence of a positive linear association (correlation) between y and x, which is a good for ratio estimation. The following data table gives $(y_i - rx_i)$ column along with x_i and y_i column, where

$$r = \frac{\sum y_i}{\sum x_i} = \frac{142}{6.7} = 21.194$$

An estimate of τ_y is, using Equation (6.4),

$$\hat{\tau}_y = r\tau_x = 21.194(75) = 1589.55$$

The standard deviation s_r is simply the sample standard deviation of the values for $(y_i - rx_i)$. Then the estimated variance of $\hat{\tau}_y$ is, from Equation (6.5),

$$\hat{V}(\hat{\tau}_y) = \tau_x^2\left(\frac{N-n}{nN}\right)\left(\frac{1}{\mu_x^2}\right)s_r^2 = (N\mu_x)^2\left(\frac{N-n}{nN}\right)\left(\frac{1}{\mu_x^2}\right)s_r^2 = N\left(\frac{N-n}{n}\right)s_r^2$$

$$B = 2\sqrt{\hat{V}(\hat{\tau}_y)} = 2\sqrt{\frac{250(250-12)}{12}}(1.323) = 186.32$$

Data for Exercise 6.1

Tree	x_i	y_i	$y_i - rx_i$
1	0.3	6	-0.35821
2	0.5	9	-1.59701
3	0.4	7	-1.47761
4	0.9	19	-0.07463
5	0.7	15	0.16418
6	0.2	5	0.76119
7	0.6	12	-0.71642
8	0.5	9	-1.59701
9	0.8	20	3.04478
10	0.4	9	0.52239
11	0.8	18	1.04478
12	0.6	13	0.28358

	n	sum	stdev
x_i	12	6.7	0.2151
y_i	12	142	5.1845
$y_i - rx_i$	12	0	1.323

Scatter plot of x and y

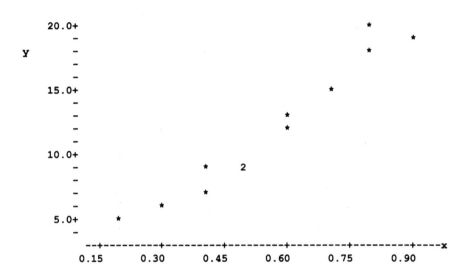

6.3 Data for Exercise 6.3

student	x_i	y_i	$y_i - rx_i$
1	25100	3800	117.33
2	32200	5100	375.62
3	29600	4200	-142.91
4	35000	6200	1064.80
5	34400	5800	752.83
6	26500	4100	211.92
7	28700	3900	-310.86
8	28200	3600	-537.50
9	34600	3800	-1276.51
10	32700	4100	-697.74
11	31500	4500	-121.68
12	30600	5100	610.37
13	27700	4200	135.86
14	28500	4000	-181.52

	n	sum	stdev
x_i	14	425300	3127.3
y_i	14	62400	792.96
$y_i - rx_i$	14	0	611.52

$$r = \frac{\sum y_i}{\sum x_i} = \frac{62400}{425300} = .147$$

$$2\sqrt{\hat{V}(r)} = 2\sqrt{\left(\frac{N-n}{nN}\right)\left(\frac{1}{\bar{x}^2}\right)s_r^2}$$

$$= 2\sqrt{\frac{150-14}{14(150)}\frac{611.52}{(425300/14)}} = .0102$$

6.5 $\quad \hat{\mu}_y = r\mu_x = r\frac{\tau_x}{N} = \frac{15422}{13547}\left(\frac{128200}{123}\right) = 1186.53$

$$\hat{V}(\hat{\mu}_y) = \mu_x^2 \hat{V}(r) = \mu_x^2\left(\frac{N-n}{nN}\right)\frac{1}{\mu_x^2}s_r^2 = \frac{N-n}{nN}s_r^2$$

$$B = 2\sqrt{\hat{V}(\hat{\mu}_y)} = 2\sqrt{\frac{123-13}{13(123)}}(113.97) = 59.79$$

6.7 $\quad x_i =$ original weight, $y_i =$ current weight

$$\sum y_i = 39.9, \quad \sum y_i^2 = 159.43, \quad \sum x_i = 29.7, \quad \sum x_i^2 = 88.43, \quad \sum x_i y_i = 118.67$$

$$r = \frac{\sum y_i}{\sum x_i} = \frac{39.9}{29.7} = 1.34$$

$$\hat{\mu}_y = r\mu_x = 1.34(3.1) = 4.16$$

$$\hat{V}(\hat{\mu}_y) = \left(\frac{N-n}{nN}\right)\frac{1}{n-1}[\sum y_i^2 - 2r\sum x_i y_i + r^2\sum x_i^2]$$

$$= \frac{100-10}{10(100)}\frac{1}{9}[159.34 - 2(1.34)(118.67) + 1.34^2(88.43)]$$

$$B = 2\sqrt{\hat{V}(\hat{\mu}_y)} = .085$$

6.9 x_i = photo count, y_i = ground count

	n	mean	stdev
x_i	10	23.4	11.4717
y_i	10	30.6	14.8489
$y_i - rx_i$	10	0.00	3.4751

$$r = \frac{\sum y_i}{\sum x_i} = \frac{\bar{y}}{\bar{x}} = \frac{30.6}{23.4} = 1.31$$

$$\hat{\tau}_y = r\tau_x = 1.31(4200) = 5492.31$$

$$\hat{V}(\hat{\tau}_y) = \tau_x^2 \left(\frac{N-n}{nN}\right)\frac{1}{\mu_x^2}s_r^2 = N^2\left(\frac{N-n}{nN}\right)s_r^2$$

$$B = 2\sqrt{\hat{V}(\hat{\tau}_y)} = 2(200)\sqrt{\frac{200-10}{10(200)}}3.4751 = 428.44$$

6.11

	n	sum	stdev
x_i	12	10103	242.4277
y_i	12	11458	215.9536
$y_i - rx_i$	12	0	94.0672

$$r = \frac{\sum y_i}{\sum x_i} = \frac{11458}{10103} = 1.134$$

$$\hat{\mu}_y = r\mu_x = 1.134(880) = 997.92$$

$$B = 2\sqrt{\hat{V}(\hat{\mu}_y)} = 2\sqrt{\left(\frac{N-n}{nN}\right)s_r^2} = 2\sqrt{\frac{500-12}{12(500)}}(94.067) = 53.65$$

6.13 $B = 3800, N = 452,$ $\hat{\sigma} = 15.5544$
From Equation (6.23),

$$n = \frac{N\hat{\sigma}^2}{ND+\hat{\sigma}^2} \quad \text{where} \quad D = \frac{B^2}{4N^2}$$

$$n = \frac{452(15.5544)^2}{\dfrac{(3800)^2}{4(452)} + (15.5544)^2} = 13.29 \approx 14$$

6.15 $\hat{\tau}_{yL} = N\hat{\mu}_{yL} = N[\bar{y} + b(\mu_x - \bar{x})]$ where $b = \dfrac{\sum x_i y_i - n\bar{x}\bar{y}}{\sum x_i^2 - n\bar{x}^2}$

$$\hat{V}(\hat{\tau}_{yL}) = \hat{V}(N\hat{\mu}_{yL}) = N^2\hat{V}(\hat{\mu}_{yL})$$

6.17 $\sum y_i = 694.6$, $\sum y_i^2 = 26841.6$

$\sum x_i = 831.6$, $\sum x_i^2 = 38495.09$, $\sum x_i y_i = 32125.1125$

$$r = \frac{\sum y_i}{\sum x_i} = \frac{694.6}{831.6} = .84$$

$$\hat{V}(r) = \left(\frac{N-n}{nN}\right)\frac{1}{\bar{x}^2}\frac{1}{n-1}[\sum y_i^2 - 2r\sum x_i y_i + r^2 \sum x_i^2]$$

$$= \left(\frac{64-18}{18(64)}\right)\frac{1}{(831.6/18)^2}\frac{1}{17}[26841.6 - 2(.84)(32125.1125$$

$$+(.84)^2(38495.09)]$$

$$B = 2\sqrt{\hat{V}(r)} = .012$$

6.21 (a) y = weekend method, x = traditional method

$$\sum y_i = 96, \quad \sum y_i^2 = 1558$$
$$\sum x_i = 92, \quad \sum x_i^2 = 1426, \quad \sum x_i y_i = 1486$$

$$r = \frac{\sum y_i}{\sum x_i} = \frac{96}{92} = 1.043$$

$$\hat{V}(r) = \frac{1}{n}\frac{1}{\mu_x^2}\frac{1}{n-1}[\sum y_i^2 - 2r\sum x_i y_i + r^2 \sum x_i^2]$$

$$= \frac{1}{6}\frac{1}{(92/6)^2}\frac{1}{5}[1558 - 2(1.043)(1468) + (1.043)^2(1426)]$$

(Use $\bar{x}$ to estimate μ_x and ignore the fpc, assuming N is large)

$$B = 2\sqrt{\hat{V}(r)} = .0733$$

(b) y = purchase method, x = traditional method

$$\sum y_i = 80, \quad \sum y_i^2 = 1150, \quad \sum x_i = 92, \quad \sum x_i^2 = 1426, \quad \sum x_i y_i = 1250$$

$$r = \frac{\sum y_i}{\sum x_i} = \frac{80}{92} = .87$$

$$\hat{V}(r) = \left(\frac{1}{n}\right)\frac{1}{\mu_x^2}\frac{1}{n-1}\left[\sum y_i^2 - 2r\sum y_i x_i + r^2\sum x_i^2\right]$$

$$= \left(\frac{1}{6}\right)\frac{1}{(92/6)^2}\frac{1}{5}\left[1150 - 2(.87)(1250) + (.87)^2(1426)\right]$$

$$B = 2\sqrt{\hat{V}(r)} = .176$$

(Use $\bar{x}$ to estimate μ_x, and ignore the fpc, assuming N is large)

6.23 **(a)** $y = 1989$ incomes, $\quad x = 1980$ incomes

$$\sum y_i = 327, \quad \sum y_i^2 = 22131$$

$$\sum x_i = 246, \quad \sum x_i^2 = 12602, \quad \sum x_i y_i = 16243$$

$$r = \frac{\sum y_i}{\sum x_i} = \frac{327}{246} = 1.329$$

$$\hat{\tau}_y = r\tau_x = (1.329)(674) = 895.75$$

$$\hat{V}(\hat{\tau}_y) = N^2\left(\frac{N-n}{nN}\right)\frac{1}{n-1}[\sum y_i^2 - 2r\sum x_i y_i + r^2\sum x_i^2]$$

$$= (19)^2\left(\frac{19-6}{6(19)}\right)\frac{1}{5}\left[22131 - 2(1.329)(16243) + (1.329)^2(12062)\right]$$

$$B = 2\sqrt{\hat{V}(\hat{\tau}_y)} = 92.78$$

(b) $\quad \hat{\tau}_{yL} = N\hat{\mu}_{yL} = N[\bar{y} + b(\mu_x - \bar{x})] \qquad$ where $\quad b = \dfrac{\sum x_i y_i - n\bar{x}\bar{y}}{\sum x_i^2 - n\bar{x}^2}$

$$b = \frac{16243 - 6(246/6)(327/6)}{12062 - 6(246/6)^2} = 1.435$$

$$\hat{\tau}_{yL} = 19\left[\frac{327}{6} + 1.435\left(\frac{674}{19} - \frac{246}{6}\right)\right] = 884.802$$

$$\hat{V}(\hat{\mu}_{yL}) = \frac{N-n}{nN}\frac{1}{n-2}\left[\sum(y_i - \bar{y})^2 - b^2\sum(x_i - \bar{x})^2\right]$$

$$= \frac{N-n}{nN}\frac{1}{n-2}\left[\left(\sum y_i^2 - n\bar{y}^2\right) - b^2\left(\sum x_i^2 - n\bar{x}^2\right)\right]$$

$$= \frac{19-6}{6(19)}\frac{1}{4}\left[\left(22131 - 6(54.5)^2\right) - (1.435)^2\left(12062 - 6(41)^2\right)\right]$$

31

$$B = 2\sqrt{\hat{V}(\hat{\tau}_{yL})} = 2\sqrt{\hat{V}(N\hat{\mu}_{yL})} = 2N\sqrt{\hat{V}(\hat{\mu}_{yL})} = 99.23$$

(c) $d_i = y_i - x_i$ = increase from 1980 to 1989

Industry	d_i
Lumber and wood products	5
Electric and electronic equipment	28
Motor vehicles and equipment	11
Food and kindred products	10
Textile mill products	1
Chemicals and allied products	26

$$\sum d_i = 81, \quad \sum d_i^2 = 1707$$

$$\hat{\tau}_{yD} = N\hat{\mu}_{yD} = N(\mu_x + \bar{d}) = 19\left(\frac{624}{19} + \frac{81}{6}\right) = 930.5$$

$$\hat{V}(\hat{\mu}_{yD}) = \left(\frac{N-n}{nN}\right)\frac{1}{n-1}\left[\sum d_i^2 - n\bar{d}^2\right]$$

$$= \left(\frac{19-6}{19(6)}\right)\frac{1}{5}(1707 - 6(81/6)^2)$$

$$B = 2\sqrt{\hat{V}(\hat{\tau}_{yD})} = 2N\sqrt{\hat{V}(\hat{\mu}_{yD})} = 142.143$$

6.25   Brand I $\sum y_i = 1215, \quad \sum y_i^2 = 279825$

                 $\sum x_i = 1158, \quad \sum x_i^2 = 249154, \quad \sum x_i y_i = 263670$

   Brand II $\sum y_i = 1090, \quad \sum y_i^2 = 149950$

                 $\sum x_i = 1027, \quad \sum x_i^2 = 131497, \quad \sum x_i y_i = 140210$

Use the combined estimate $\hat{\tau}_{yRC}$.

$$\hat{\tau}_{yRC} = N\hat{\mu}_{yRC} = N\frac{\bar{y}_{st}}{\bar{x}_{st}}\mu_x = \frac{\bar{y}_{st}}{\bar{x}_{st}}\tau_x$$

$$\bar{y}_{st} = \frac{1}{N}(N_1\bar{y}_1 + N_2\bar{y}_2) = \frac{1}{300}\left[120\frac{1215}{6} + 180\frac{1090}{9}\right] = 153.67$$

$$\bar{x}_{st} = \frac{1}{N}(N_1\bar{x}_1 + N_2\bar{x}_2) = \frac{1}{300}\left[120\frac{1158}{6} + 180\frac{1027}{9}\right] = 145.67$$

$$\hat{\tau}_{yRC} = \frac{153.67}{145.67}(24500 + 21200) = 48209.84$$

$$\hat{V}(\hat{\tau}_{yRC}) = N^2 \hat{V}(\hat{\mu}_{yRC})$$

$$= N^2 \left[\left(\frac{N_1}{N}\right)^2 \left(\frac{N_1 - n_1}{N_1 n_1}\right) \left(\frac{1}{n_1 - 1}\right) \sum \left[(y_{i1} - \bar{y}_1) - r_c(x_{i1} - \bar{x}_1)\right] \right]$$

$$+ N^2 \left[\left(\frac{N_2}{N}\right)^2 \left(\frac{N_2 - n_2}{N_2 n_2}\right) \left(\frac{1}{n_2 - 1}\right) \sum \left[(y_{i2} - \bar{y}_2) - r_c(x_{i2} - \bar{x}_2)\right] \right]$$

$$= 300^2 \left[\left(\frac{120}{300}\right)^2 \frac{120 - 6}{120(6)} \frac{1}{5}(788.81) + \left(\frac{180}{300}\right)^2 \frac{180 - 9}{180(9)} \frac{1}{8}(461.75) \right]$$

$$= 557095.07$$

where

$$r_c = \frac{\bar{y}_{st}}{\bar{x}_{st}}$$

For Brand I, $\quad \sum \left[(y_{i1} - \bar{y}_1) - r_c(x_{i1} - \bar{x}_1)\right]^2 = 788.81$

For Brand II, $\quad \sum \left[(y_{i2} - \bar{y}_2) - r_c(x_{i2} - \bar{x}_2)\right]^2 = 461.75$

6.27 $\quad b = \dfrac{\sum x_i y_i - n\bar{x}\bar{y}}{\sum x_i^2 - n\bar{x}^2} = \dfrac{3090.93 - 11(181.2/11)(185.3/11)}{3027.06 - 11(181.2/11)^2} = .9131$

$$\hat{V}(\hat{\mu}_{yRL}) = \left(\frac{N-n}{nN}\right)\frac{1}{n-2}\left[\left(\sum y_i^2 - n\bar{y}^2\right) - b^2\left(\sum x_i^2 - n\bar{x}^2\right)\right]$$

$$= \frac{763-11}{11(763)}\frac{1}{9}\left[\left(3158.19 - 11\left(\frac{185.3}{11}\right)^2\right) - (.9131)^2\left(3027.06 - 11\left(\frac{181.2}{11}\right)^2\right)\right]$$

$$= .01536$$

$$\hat{V}(\hat{\mu}_y) = \left(\frac{N-n}{nN}\right)\frac{1}{n-1}\left[\sum y_i^2 - 2r\sum x_i y_i + r^2 \sum x_i^2\right]$$

$$= \left(\frac{763-11}{111(763)}\right)\frac{1}{10}\left[3158.19 - 2(1.02)(3090.93) + 1.02^2(3027.06)\right] = .01836$$

$$\hat{V}(\hat{\mu}_{yD}) = \left(\frac{N-n}{nN}\right)\frac{1}{n-1}\left[\sum d_i^2 - n\bar{d}^2\right]$$

$$= \left(\frac{N-n}{nN}\right)\frac{1}{n-1}\left[\sum (y_i - x_i)^2 - n(\bar{y}-\bar{x})^2\right]$$

$$= \left(\frac{N-n}{nN}\right)\frac{1}{n-1}\left[\left(\sum y_i^2 - 2\sum x_i y_i + \sum x_i^2\right) - n(\bar{y}-\bar{x})^2\right]$$

$$= \left(\frac{763-11}{111(763)}\right)\frac{1}{10}\left[3158.19 - 2(3090.93) + 3027.06 - 11\left(\frac{185.3}{11} - \frac{181.2}{11}\right)^2\right]$$

$$= .01668$$

(a) $\quad \text{RE}(\hat{\mu}_{yRL} / \hat{\mu}_y) = \hat{V}(\hat{\mu}_y)/\hat{V}(\hat{\mu}_{yRL}) = .01836 / .01536 = 1.195$

(b) $\quad \text{RE}(\hat{\mu}_{yRL} / \hat{\mu}_{yD}) = \hat{V}(\hat{\mu}_{yD})/\hat{V}(\hat{\mu}_{yRL}) = .01668 / .01536 = 1.086$

(c) $\quad \text{RE}(\hat{\mu}_y / \hat{\mu}_{yD}) = \hat{V}(\hat{\mu}_{yD})/\hat{V}(\hat{\mu}_y) = .01668 / .01836 = .908$

6.33 $\quad y = \text{weight}, \quad x = \text{length}$

$$\sum y_i = 2648, \quad \sum y_i^2 = 698878$$
$$\sum x_i = 2046, \quad \sum x_i^2 = 184198, \quad \sum x_i y_i = 283373$$

$$\hat{\mu}_{yL} = \bar{y} + b(\mu_x - \bar{x}) = \frac{2648}{24} + 5.895\left(100 - \frac{2046}{24}\right) = 197.28$$

where

$$b = \frac{\sum x_i y_i - n\bar{x}\bar{y}}{\sum x_i^2 - n\bar{x}^2} = \frac{283373 - 24(2046/24)(2648/24)}{184198 - 24(2046/24)^2} = 5.895$$

CHAPTER 7
SYSTEMATIC SAMPLING

7.1   Systematic sampling is better than sample random sampling since population is ordered and N is large.

7.3   **(a)** $N = 40$, $k = 10$, $n = 4$

sample	sample elements	$\hat{p}$
1	1 11 21 31	0.75
2	2 12 22 32	1.00
3	3 13 23 33	0.75
4	4 14 24 34	0.75
5	5 15 25 35	0.25
6	6 16 26 36	0.00
7	7 17 27 37	0.00
8	8 18 28 38	0.00
9	9 19 29 39	0.25
10	10 20 30 40	0.25

where $\hat{p}$ is the proportion of deliquent accounts in the sample

The probability distribution of $\hat{p}$ is

$\hat{p}$	0.00	0.25	0.75	1.00
$p(\hat{p})$	0.3	0.3	0.3	0.1

$E(\hat{p}) = \sum \hat{p}p(\hat{p}) = 0(.3) + .25(.3) + .75(.3) + 1(.1) = .4$

$E(\hat{p}^2) = \sum \hat{p}^2 p(\hat{p}) = (0)^2(.3) + (.25)^2(.3) + (.75)^2(.3) + (1)^2(.1) = .2875$

$V(\hat{p}) = E(\hat{p}^2) - \left(E(\hat{p})\right)^2 = .2875 - .16 = .1275$

(b) $N = 40$, $k = 5$, $n = 8$

sample	sample elements								$\hat{p}$
1	1	6	11	16	21	26	31	36	0.375
2	2	7	12	17	22	27	32	37	0.500
3	3	8	13	18	23	28	33	38	0.375
4	4	9	14	19	24	29	34	39	0.500
5	5	10	15	30	25	30	35	40	0.250

where $\hat{p}$ is the proportion of deliquent accounts in the sample

The probability distribution of $\hat{p}$ is

$\hat{p}$	0.25	0.375	0.50
$p(\hat{p})$	0.2	0.4	0.4

$E(\hat{p}) = \sum \hat{p}p(\hat{p}) = .25(.2) + .375(.4) + .5(.4) = .4$

$E(\hat{p}^2) = \sum \hat{p}^2 p(\hat{p}) = (.25)^2(.2) + (.375)^2(.4) + (.5)^2(.4) = .16875$

$V(\hat{p}) = E(\hat{p}^2) - \left(E(\hat{p})\right)^2 = .18675 - (.4)^2 = .00875$

7.5 $N = 2000$, $\hat{p}_{sy} = .66$, $\hat{q}_{sy} = .34$, $D = B^2/4 = (.01)^2/4 = (0.005)^2$

$n = \dfrac{Npq}{(N-1)D+pq} \approx \dfrac{N\hat{p}_{sy}\hat{q}_{sy}}{(N-1)D+\hat{p}_{sy}\hat{q}_{sy}}$

$= \dfrac{2000(.66)(.34)}{1999(.005)^2 + (.66)(.34)} = 1635.72 \approx 1636$

7.7 $N = 1800$ $s^2 = .0062$ $D = B^2/4 = .03^2/4$

$n = \dfrac{N\sigma^2}{(N-1)D+\sigma^2} \approx \dfrac{Ns^2}{(N-1)D+s^2} = 27.02 \approx 28$

7.9 $\hat{p}_{sy} = \dfrac{1}{n}\sum y_i = \dfrac{324}{400} = .81$

$\hat{V}(\hat{p}_{sy}) = \dfrac{\hat{p}_{sy}\hat{q}_{sy}}{n-1}\left(\dfrac{N-n}{N}\right) = \dfrac{.81(.19)}{399}\left(\dfrac{2800-400}{2800}\right)$

$B = 2\sqrt{\hat{V}(\hat{p}_{sy})} = .036$

7.11 $N = 4500$, $n = 30$

$$\sum y_i = 850 \quad \sum y_i^2 = 33904 \quad s^2 = 338.64$$

$$\hat{\tau} = N\bar{y}_{sy} = N\bar{y} = 4500(850/30) = 127500$$

$$\hat{V}(\hat{\tau}) = N^2 \frac{s^2}{n}\left(\frac{N-n}{N}\right) = (4500)^2 \frac{338.64}{30}\left(\frac{4500-30}{4500}\right)$$

$$B = 2\sqrt{\hat{V}(\hat{\tau})} = 30137.06$$

7.13

		n	mean	median	stdev
Payroll	P_{91} - P_{87}	16	315.31	251.0	385.79

For estimating the mean difference, with the estimator denoted by $\bar{x}_{sy}$, we have

$$\bar{x}_{sy} = 315.31$$

$$B = 2\sqrt{\hat{V}(\bar{x}_{sy})} = 2\sqrt{\left(\frac{N-n}{N}\right)\frac{s^2}{n}} = 2\sqrt{\frac{79-16}{79}\frac{385.79}{4}} = 172.26$$

Thus, $\bar{x}_{sy} \pm B$ yields an interval of (143.05, 487.57). We can conclude that there was a significant increase in average amount of payroll per industry group over the period from 1987 to 1991. Note that for the numeber of employees, there was a significant decrease over the given period.

7.15 $N = 180 \quad n = 18$

$$\sum y_i = 4868 \quad \sum y_i^2 = 1321450 \quad s^2 = 289.79$$

$$\hat{\tau} = N\bar{y}_{sy} = N\bar{y} = 180(4868/18) = 48680$$

$$B = 2\sqrt{\hat{V}(\hat{\tau})} = 2N\sqrt{\frac{s^2}{n}\left(\frac{N-n}{N}\right)} = 2(180)\sqrt{\frac{289.79}{18}\left(\frac{180-18}{180}\right)} = 1370.34$$

7.17 $N = 650$, $n = 65$, $\sum y_i = 48$

$$\hat{p}_{sy} = \frac{1}{n}\sum y_i = \frac{48}{65} = .738$$

$$B = 2\sqrt{\hat{V}(\hat{p}_{sy})} = 2\sqrt{\frac{\hat{p}_{sy}\hat{q}_{sy}}{n-1}\left(\frac{N-n}{N}\right)} = 2\sqrt{\frac{.74(.26)}{64}\left(\frac{650-65}{650}\right)} = .104$$

7.21 $N =$ number of years in the study (1950 - 1990) $= 41$
$n =$ number of births $= 9$

$$\sum y_i = 6932 \quad \sum y_i^2 = 6459251 \quad s^2 = 140009$$

$$\hat{\tau} = N\bar{y}_{sy} = N\bar{y} = 41(6932/9) = 31579.11$$

$$B = 2\sqrt{\hat{V}(\hat{\tau})} = 2\sqrt{N^2\frac{s^2}{n}\left(\frac{N-n}{N}\right)} = 2\sqrt{(41)^2\frac{140009}{9}\left(\frac{41-9}{41}\right)} = 9035.53$$

The following plot of year vs. rate shows a definite increasing trend of number of divorces as year advances. The variance of approximation from simple random sampling will overestimate the true variance.

MTB > plot 'divorces' 'year'

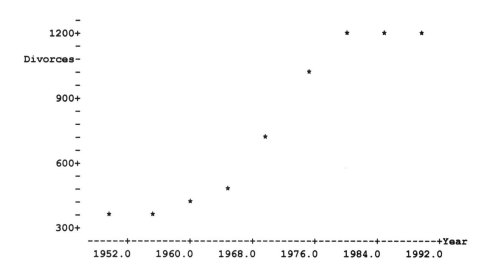

7.27 **(a)** For exercise 7.6

Successive Differences (d_i)

.09	.12	.05	.04	.06	.10	.03
.04	.01	.03	.08	.04	.04	.05
.18	.03	.02	.16	.08	.01	.01
.33	.14	.05	.05	.04	.09	.16
.13	.06	.07	.11	.01	.22	.01

$$\sum d_i^2 = .3766$$

$$\hat{V}_d(\bar{y}_{sy}) = \frac{N-n}{nN}\frac{1}{2(n-1)}\sum d_i^2 = \frac{1800-36}{36(1800)}\frac{1}{2(35)}(.3766) = .000146$$

$$\hat{V}(\bar{y}_{sy}) = \frac{s^2}{n}\left(\frac{N-n}{N}\right) = \frac{.0062}{35}\left(\frac{1800-36}{1800}\right) = .000168$$

7.27 **(b)** For exercise 7.11

Successive Differences (d_i)

8	9	8	18	6	7
12	11	8	15	9	1
52	12	2	19	87	17
34	34	8	2	72	9
3	22	9	41	1	

$$\sum d_i^2 = 22226$$

$$\hat{V}_d(\hat{\tau}_{sy}) = N^2 \hat{V}_d(\bar{y}_{sy}) = N^2 \frac{N-n}{nN}\frac{1}{2(n-1)}\sum d_i^2$$

$$= (4500)^2 \frac{4500-30}{30(4500)}\frac{1}{2(29)}(22226) = 256940224.1$$

$$\hat{V}(\hat{\tau}_{sy}) = N^2 \frac{s^2}{n}\left(\frac{N-n}{N}\right) = (4500)^2 \frac{338.64}{30}\left(\frac{4500-30}{4500}\right) = 227060586.2$$

(c) For exercise 7.16

Successive Differences (d_i)

310	220	340	180	60
130	370	470	300	250
360	70	160	100	370
60	360	820	240	330

$$\sum d_i^2 = 2102400$$

$$\hat{V}_d(\bar{y}_{sy}) = \frac{N-n}{nN}\frac{1}{2(n-1)}\sum d_i^2 = \frac{520-21}{21(520)}\frac{1}{2(20)}(2102400) = 2401.78$$

$$\hat{V}(\bar{y}_{sy}) = \frac{s^2}{n}\left(\frac{N-n}{N}\right) = \frac{64686.19}{21}\left(\frac{520-21}{520}\right) = 2955.90$$

(d) For exercise 7.20 (a),

Successive Differences (d_i)

465	161	498	29	587	468	149	397

$$\sum d_i^2 = 1234,394$$

$$\hat{V}_d(\hat{\tau}_{sy}) = N^2\hat{V}_d(\bar{y}_{sy}) = N^2\left(\frac{N-n}{nN}\right)\frac{1}{2(n-1)}\sum d_i^2$$

$$= (41)^2\left(\frac{41-9}{9(41)}\right)\frac{1}{2(8)}(1234394) = 11246701$$

$$\hat{V}(\hat{\tau}_{sy}) = N^2\frac{s^2}{n}\left(\frac{N-n}{N}\right) = (41)^2\frac{115959}{9}\left(\frac{41-9}{41}\right) = 16904245$$

For exercise 7.20 (b),

Successive Differences (d_i)

 .9   1.3   4.3   1.0   3.8   1.3   .1   .9

$$\sum d_i^2 = 38.94$$

$$\hat{V}_d(\bar{y}_{sy}) = \frac{N-n}{nN}\frac{1}{2(n-1)}\sum d_i^2 = \frac{41-9}{9(41)}\frac{1}{2(8)}(38.94) = .21$$

$$\hat{V}(\bar{y}_{sy}) = \frac{s^2}{n}\left(\frac{N-n}{N}\right) = \frac{16.0461}{9}\left(\frac{41-9}{41}\right) = 1.39$$

7.27 **(e)**   For exercise 7.21 ,

Successive Differences (d_i)

 8   16   86   229   328   153   10   15

$$\sum d_i^2 = 191475$$

$$\hat{V}_d(\hat{\tau}_{sy}) = N^2\left(\frac{N-n}{nN}\right)\frac{1}{2(n-1)}\sum d_i^2$$

$$= (41)^2\left(\frac{41-9}{9(41)}\right)\frac{1}{2(8)}(191475) = 1744550$$

$$\hat{V}(\hat{\tau}_{sy}) = N^2\frac{s^2}{n}\left(\frac{N-n}{N}\right) = (41)^2\frac{140009}{9}\left(\frac{41-9}{41}\right) = 2.04102 \times 10^7$$

CHAPTER 8
CLUSTER SAMPLING

8.2   The following able is the data for exercise 8.2 and the basic summary statistics.
$(y_i - \bar{y}m_i)$ column is established since the basic component of the estimated
variance is the sample variance of these differences, where

$$\bar{y} = \frac{\sum y_i}{\sum m_i} = \frac{2565}{130} = 19.73$$

which is the estimate of the population mean μ, by using Equation (8.1).

Cluster	m_i	y_i	$y_i - \bar{y}m_i$
1	3	50	-9.1923
2	7	110	-28.1154
3	11	230	12.9615
4	9	140	-37.5769
5	2	60	20.5385
6	12	280	43.2308
7	14	240	-36.2308
8	3	45	-14.1923
9	5	60	-38.6539
10	9	230	52.4231
11	8	140	-17.8462
12	6	130	11.6154
13	3	70	10.8077
14	2	50	10.5385
15	1	10	-9.7308
16	4	60	-18.9231
17	12	280	43.2308
18	6	150	31.6154
19	5	110	11.3461
20	8	120	-37.8462

Summary statistics

	n	mean	sum	stdev	
m	20	6.5	130	3.791	$N = 96$
y_i	20	128.3	2565	83.118	
$y_i - \bar{y}m_i$	20	0	0	29.079	

Since M is not known, the $\overline{M}$ appearing in Equation (8.2) must be estimated by $\overline{m}$, where

$$\overline{m} = \frac{\sum m_i}{n} = \frac{130}{20} = 6.5$$

The from Equation (8.2), the estimated variance of population mean is,

$$\hat{V}(\overline{y}) = \left(\frac{N-n}{Nn\overline{M}^2}\right)\frac{1}{n-1}\sum(y_i - \overline{y}m_i)^2 = \left(\frac{N-n}{Nn\overline{M}^2}\right)s_r^2$$

$$= \left(\frac{96-20}{96(20)(6.5)^2}\right)(29.08)^2$$

with

$$B = 2\sqrt{\hat{V}(\overline{y})} = 1.78$$

A plot of y versus m for the data is given in the following. Note that the linearity is fairly strong.

```
MTB > plot 'y' 'm'

        -
   300+
        -                                                              2
 Y      -
        -                                          *         *              *
        -
   200+
        -
        -                              *
        -                              *              *    *
        -                    *              *    *
   100+
        -            *
        -       2    *    *    *
        -            *
        -  *
    0+
        --------+---------+---------+---------+---------+---------+------m
            2.5       5.0       7.5      10.0      12.5
```

8.3   By Equation (8.7), the estimate of the population total τ is,

$$\hat{\tau} = N\overline{y}_t = N\frac{\sum y_i}{n} = 96\frac{2565}{20} = 12312$$

The estimated variance is, using Equation (8.8)

$$\hat{V}(\hat{\tau}) = \hat{V}(N\bar{y}_t) = N^2 \left(\frac{N-n}{Nn} \right) s_t^2 = (96)^2 \left(\frac{96-20}{96(20)} \right) (83.118)^2$$

where

$$s_t^2 = \frac{1}{n-1} \sum (y_t - \bar{y}_t)^2 = 83.118$$

(from table of Summary statistics of Exercise 8.2)

A bound on the error of estimation of τ is

$$B = 2\sqrt{\hat{V}(N\bar{y}_t)} = 3175.06$$

8.5 From equation (8.2),

$$n = \frac{N\sigma_r^2}{ND + \sigma_r^2} \quad \text{where} \quad D = \frac{B^2 \overline{M}^2}{4}$$

Here $B = 2$, $\overline{M} = M / N = 710 / 96 = 7.4$ and s_r^2 is used to estimate σ_r^2.
Then

$$n = \frac{96(29.079)^2}{96(7.4)^2 + (29.079)^2} = 13.3 \approx 14$$

8.8 The following table is the data for Exercise 8.8 and the basic summary statistics.
$(a_i - \bar{a}m_i)$ column is established since the basic component of the estimated
variance is the sample variance of these differences.

Cluster	m_i	a_i	$a_i - \bar{a}m_i$
1	51	42	5.83535
2	62	53	9.03513
3	49	40	5.25357
4	73	45	-6.76509
5	101	63	-8.62019
6	48	31	-3.03732
7	65	38	-8.09221
8	49	30	-4.74643
9	73	54	2.23491
10	61	45	1.74424
11	58	51	9.87157
12	52	29	-7.87376
13	65	46	-0.09221
14	49	37	2.25357
15	55	42	2.99890

Summary statistics

	n	mean	sum	stdev
m	15	60.73	911	14.007
a_i	15	43.07	646	9.5728
$a_i - \bar{a}m_i$	15	0	0	6.2230

$N = 87$

By Equation (8.16), the estimate of the population proportion p is,

$$\hat{p} = \frac{\sum a_i}{\sum m_i} = \frac{646}{911} = .709$$

Since M is not known, the $\bar{M}$ appearing in Equation (8.17) must be estimated by $\bar{m}$, where

$$\bar{m} = \frac{\sum m_i}{n} = \frac{911}{15} = 60.73$$

The from Equation (8.17), the estimated variance of population proportion is,

$$\hat{V}(\hat{p}) = \left(\frac{N-n}{Nn\bar{M}^2}\right)\frac{1}{n-1}\sum(a_i - \bar{a}m_i)^2 = \left(\frac{N-n}{Nn\bar{M}^2}\right)s_p^2$$

$$= \left(\frac{87-15}{87(15)(60.73)^2}\right)(6.223)^2$$

with $B = 2\sqrt{\hat{V}(\hat{p})} = 0.048$

A plot of a versus m for the data is given in the following. Note that the linearity is fairly strong.

```
MTB > plot 'a' 'm'
```

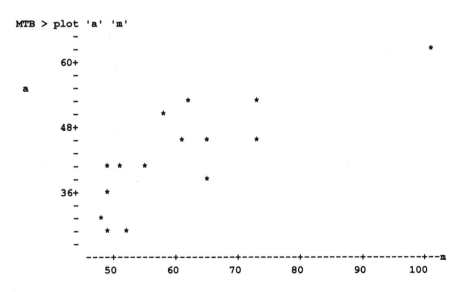

44

8.9   From section 8.7,

$$n = \frac{N\sigma_p^2}{ND + \sigma_p^2} \qquad \text{where} \qquad D = \frac{B^2 \overline{M}^2}{4}$$

Here $N = 87$, $B = .08$, $\overline{m}$ is used to estimate $\overline{M}$ for D and s_p^2 is used to estimate σ_p^2. These values are obtained from Exercise 8.8.

Then

$$n = \frac{87(6.223)^2}{87\left[(.08)^2 (60.73)^2 / 4\right] + (6.223)^2} = 6.1 \approx 7$$

8.10   Summary statistics

	n	mean	sum	stdev	
m	20	65.15	1303	8.2096	$N = 60$
y_i	20	2617.0	52340	316.21	
$y_i - \bar{y}m_i$	20	0	0	114.27	

By Equation (8.1), the estimate of the population mean μ is,

$$\bar{y} = \frac{\sum y_i}{\sum m_i} = \frac{52340}{1303} = 40.17$$

Since M is not known, the $\overline{M}$ appearing in Equation (8.2) must be estimated by $\overline{m}$, where

$$\overline{m} = \frac{\sum m_i}{n} = \frac{1303}{20} = 65.15$$

The from Equation (8.2), a bound on the error of estimation is,

$$B = 2\sqrt{\hat{V}(\bar{y})} = 2\sqrt{\left(\frac{N-n}{Nn\overline{M}^2}\right)s_r^2} = 2\sqrt{\frac{N-n}{Nn}}\frac{s_r}{\overline{M}}$$

$$= 2\sqrt{\frac{60-20}{60(20)}}\frac{114.27}{65.15} = 0.64$$

8.11 By Equation (8.7), the estimate of the population total τ is,

$$\hat{\tau} = N\bar{y}_t = N\frac{\sum y_t}{n} = 60\frac{52340}{20} = 157020$$

A bound on the error of estimation is, using Equation (8.8),

$$B = 2\sqrt{\hat{V}(\hat{\tau})} = 2\sqrt{\hat{V}(N\bar{y}_t)} = 2\sqrt{N^2\left(\frac{N-n}{Nn}\right)s_t^2} = 2\sqrt{\frac{N(N-n)}{n}}s_t$$

$$= 2\sqrt{\frac{60(60-20)}{20}}(316.21) = 6927.8$$

where

$$s_t^2 = \frac{1}{n-1}\sum(y_i - \bar{y}_t)^2 = 316.21$$

(from table of Summary statistics of Exercise 8.10)

8.13

	Carton				
	1	2	3	4	5
y_i	192.0	192.1	192.0	192.5	191.7
m_i	12	12	12	12	12

$\sum y_i = 960.3 \quad \sum y_i^2 = 184435.55$

$\sum m_i = 60 \quad \sum m_i^2 = 720 \quad \sum m_i y_i = 11523.6$

$\bar{M} = 12$ boxes per carton

$$\bar{y} = \frac{\sum y_i}{\sum m_i} = \frac{960.3}{60} = 16.005$$

By the Equation (8.2), ignoring the fpc,

$$\hat{V}(\bar{y}) = \left(\frac{1}{n\bar{M}^2}\right)\frac{1}{n-1}\left[\sum y_i^2 - 2\bar{y}\sum m_i y_i + \bar{y}^2\sum m_i^2\right]$$

$$= \frac{1}{5(12)^2}\frac{1}{4}\left[184435.55 - 2(16.005)(11523.6) + (16.005)^2(720)\right]$$

(Ignoring the fpc)

$$B = 2\sqrt{\hat{V}(\bar{y})} = .0215$$

8.15 $B = .05$

$$n = \frac{N\sigma_p^2}{ND + \sigma_p^2} \qquad D = \frac{B^2 \overline{M}^2}{4}$$

Use $\overline{m}$ and s_p^2 to estimate $\overline{M}$ and σ_p^2 respectively, where

$$\overline{m} = 1548.38$$

$$s_p^2 = \frac{1}{n-1}\left[\sum y_i^2 - 2\hat{p}\sum a_i m_i + \hat{p}^2 \sum m_i^2\right]$$

$$= \frac{1}{49}\left[43752466 - 2(.57)(73792480) + .57^2(128988955)\right] = 31377.58$$

Then

$$n = \frac{497(31377.58)}{497\dfrac{.05^2(1548.38)^2}{4} + 31377.58} = 20.09 \approx 21$$

8.17 $N = 175, \ n = 25$

m_i = number of elements in cluster i = 4 tires per cab

$\overline{M}$ = average cluster size = 4

$$\sum a_i = 40 \quad \sum a_i^2 = 102$$

$$\sum m_i = 100 \quad \sum m_i^2 = 400 \quad \sum a_i m_i = 160$$

$$\hat{p} = \frac{\sum a_i}{\sum m_i} = \frac{40}{100} = .4$$

$$\hat{V}(\hat{p}) = \left(\frac{N-n}{Nn\overline{M}^2}\right)\frac{1}{n-1}\left[\sum a_i^2 - 2\hat{p}\sum a_i m_i + \hat{p}^2 \sum m_i^2\right]$$

$$= \left(\frac{175-25}{175(25)4^2}\right)\frac{1}{24}\left[102 - 2(.4)(160) + .4^2(400)\right]$$

$$B = 2\sqrt{\hat{V}(\hat{p})} = .116$$

8.19 Summary statistics

		Florida			California	
	n	mean	stdev	n	mean	stdev
m_i	8	13.625	6.05	10	14.4	8.83
y_{ti}	8	42.125	15.17	10	35.5	15.72
$y_{ti} - \overline{y}^* m_i$	8	5.54	4.8858	10	-3.17	16.946
		$N_1 = 80$			$N_2 = 140$	

An estimate of the average of sick leave per employee is then

$$\bar{y}^* = \frac{N_1\bar{y}_{t1} + N_2\bar{y}_{t2}}{N_1\bar{m}_1 + N_2\bar{m}_2} = \frac{80(42.125) + 140(35.5)}{80(13.625) + 140(14.4)} = .2685$$

The variance of $\bar{y}^*$ is

$$\hat{V}(\bar{y}^*) = \frac{1}{M^2}\left[\frac{N_1(N_1 - n_1)}{n_1}s_{r1}^2 + \frac{N_2(N_2 - n_2)}{n_2}s_{r2}^2\right]$$

$$= \frac{1}{3106^2}\left(\frac{80(80 - 8)}{8}(4.8858)^2 + \frac{140(140 - 10)}{10}(16.946)^2\right)$$

$$= .056$$

where

$$M = M_1 + M_2 = N_1\overline{M}_1 + N_2\overline{M}_2 = 80(13.625) + 140(14.4) = 3106$$

$\overline{M}_1, \overline{M}_2$ are estimated by $\bar{m}_1, \bar{m}_2$ and

$$s_{ri}^2 = \frac{1}{n_i - 1}\sum\left(y_{ti} - \bar{y}^*m_i\right)^2$$

8.21 a_i = number of defective microchips on board i
m_i = number of microchips on board i (12 per board)

$n = 10, \quad \overline{M} = 12$

$\sum a_i = 16 \quad \sum a_i^2 = 44$

$\sum m_i = 120 \quad \sum m_i^2 = 1440 \quad \sum a_i m_i = 192$

$$\hat{p} = \frac{\sum a_i}{\sum m_i} = \frac{16}{120} = .1333$$

By Equation (8.17), ignoring the fpc,

$$\hat{V}(\hat{p}) = \left(\frac{1}{n\overline{M}^2}\right)\frac{1}{n-1}\left[\sum a_i^2 - 2\hat{p}\sum a_i m_i + \hat{p}^2\sum m_i^2\right]$$

$$= \frac{1}{10(12^2)}\frac{1}{9}\left[44 - 2(.13)(192) + .13^2(1440)\right]$$

$$B = 2\sqrt{\hat{V}(\hat{p})} = .075$$

8.23 m_i = number of equipment items
a_i = number of items not properly identified

$N = 15, \quad n = 5$

$$\sum a_i = 9 \quad \sum a_i^2 = 19$$

$$\sum m_i = 98 \quad \sum m_i^2 = 2252 \quad \sum a_i m_i = 183$$

$$\hat{p} = \frac{\sum a_i}{\sum m_i} = \frac{9}{98} = .0918 \quad \overline{m} = \frac{\sum m_i}{n} = \frac{98}{5} = 19.6$$

$$\hat{V}(\hat{p}) = \left(\frac{N-n}{Nn\overline{M}^2}\right) \frac{1}{n-1} \left[\sum a_i^2 - 2\hat{p}\sum a_i m_i + \hat{p}^2 \sum m_i^2\right]$$

$$= \left(\frac{15-5}{15(5)(19.6)^2}\right) \frac{1}{4} \left[19 - 2(.0918)(183) + (.0918)^2(2252)\right]$$

($\overline{M}$ is estimated by $\overline{m}$)

$$B = 2\sqrt{\hat{V}(\hat{p})} = .039$$

8.25 $N = 15, n = 3, M = 319$

Cluster i	m_i	y_i	$\pi_i = \dfrac{m_i}{M}$	$\overline{y}_i = \dfrac{y_i}{m_i}$
6	15	2	35/319	2/15
7	18	2	18/319	2/18
11	22	2	22/319	2/22

$$\hat{\tau}_{pps} = \frac{1}{n}\sum \frac{y_i}{\pi_i} = \frac{1}{3}\left(2\frac{319}{15} + 2\frac{319}{18} + 2\frac{319}{22}\right) = 35.6593$$

$$\hat{V}(\hat{\tau}_{pps}) = \frac{1}{n(n-1)}\sum\left(\frac{y_i}{\pi_i} - \hat{\tau}_{pps}\right)^2 = 15.274$$

$$B = 2\sqrt{\hat{V}(\hat{\tau}_{pps})} = 7.82$$

8.27 $N = 10, n = 4, M = 150$

Board	m_i	y_i	$\pi_i = \dfrac{m_i}{M}$	$\overline{y}_i = \dfrac{y_i}{m_i}$
2	12	1	12/150	1/12
3	22	3	22/150	3/22
5	16	2	16/150	2/16
7	9	1	9/150	1/9

$$\hat{\mu}_{pps} = \frac{1}{Nn}\sum \frac{y_i}{\pi_i} = \frac{1}{10(4)}\left(1\frac{150}{12} + 3\frac{150}{22} + 2\frac{150}{16} + 1\frac{150}{9}\right) = 1.70928$$

$$\hat{\tau}_{pps} = N\hat{\mu}_{pps} = 10(1.70928) = 17.0928$$

$$\hat{V}(\hat{\mu}_{pps}) = \frac{1}{N^2 n(n-1)}\sum \left(\frac{y_i}{\pi_i} - \hat{\tau}_{pps}\right)^2 = .0294359$$

$$B = 2\sqrt{\hat{V}(\hat{\mu}_{pps})} = .343$$

8.29 Selecting a sample of four states with probabilities proportional to their total population is an appropriate sampling scheme for estimation total unemployment in the northeast. because the total unemployment is heavily influenced by the state of large population. But this procedure is not appropriate sampling scheme for estimating acres of forest land in the northeast. Because for estimating acres of forest land we don't need to take into account the size of population, but need to consider the size of area.

State	Size of Pop. 1980	Cumulative size
Maine	1125	1 - 1125
Mew Hampshire	921	1126 - 2046
Vermont	511	2047 - 2557
Massachusetts	5737	2558 - 8294
Rhode Island	947	8295 - 9241
Connecticut	3108	9242 - 12349
New York	17558	12350 - 29907
New Jersey	7364	29908 - 37271
Pennsylvania	11867	37272 - 49138

To choose the sample of $n = 4$ states, select 4 random numbers between 1 and 49138. Use line 6 from random number table in the Appendix. And we have 6907, 11008, 42751, and 27756. We choose Massachusetts, Connecticut, Pennsylvania, and New York.

8.31 y_i = pounds of spoiled seafood in carton i

m_i = pounds of seafood in carton i = $24 \times 5 = 120$ for all cartons

$$\sum y_i = 30 \qquad \sum a_i^2 = 230$$
$$\sum m_i = 600 \qquad \sum m_i^2 = 72000 \qquad \sum m_i y_i = 3600$$

$$\bar{y} = \frac{\sum y_i}{\sum m_i} = \frac{30}{600} = .05$$

$$\hat{\tau} = M\bar{y} = 100(120)(.05) = 600$$

$$\hat{V}(\hat{\tau}) = N^2\left(\frac{N-n}{Nn}\right)\frac{1}{n-1}\left[\sum y_i^2 - 2\bar{y}\sum m_i y_i + \bar{y}^2 \sum m_i^2\right]$$

$$= 100^2\left(\frac{100-5}{100(5)}\right)\frac{1}{4}\left[230 - 2(.05)(3600) + .05^2(72000)\right]$$

$$B = 2\sqrt{\hat{V}(\hat{\tau})} = 308.22$$

CHAPTER 9
TWO-STAGE CLUSTER SAMPLING

9.2

Plot	M_i	m_i	$\bar{y}_i$	s_i	$M_i\bar{y}_i$	$M_i(\bar{y}_i - \hat{\mu}_r)$	within
1	52	5	11.6000	1.14018	603.200	115.496	635.440
2	56	6	8.8333	1.16905	494.667	−30.553	637.778
3	60	6	5.5000	1.04881	330.000	−232.776	594.000
4	46	5	7.0000	0.70711	322.000	−109.431	188.600
5	49	5	11.6000	1.14018	568.400	108.832	560.560
6	51	5	13.4000	1.14108	683.400	205.075	609.960
7	50	5	6.8000	0.83666	340.000	−128.947	315.000
8	61	6	9.1667	0.75277	559.167	−12.948	316.861
9	60	6	8.8333	1.16905	530.000	−32.736	738.000
10	45	6	12.0000	0.63246	540.000	117.948	117.000

where *within* is defined as $M_i(M_i - m_i)s_i^2 / m_i$

Summary statistics

	n	mean	stdev
M_i	10	53.00	5.907
m_i	10	5.50	.5271
$M_i\bar{y}_i$	10	497.1	125.21
$M_i(\bar{y}_i - \hat{\mu}_r)$	10	0	135.85
within	10	471.32	216.48

Since M is unknown, we must use $\hat{\mu}_r$, given by Equation (9.7), to estimate μ.

$$\hat{\mu}_r = \frac{\sum M_i\bar{y}_i}{\sum M_i} = \frac{\sum M_i\bar{y}_i / n}{\sum M_i / n} = \frac{497.1}{53} = 9.379$$

The estimated variance of $\hat{\mu}_r$ is, from Equation (9.8),

$$\hat{V}(\hat{\mu}_r) = \frac{N-n}{Nn\overline{M}^2}s_r^2 + \frac{1}{Nn\overline{M}^2}\sum M_i^2 \frac{M_i - m_i}{M_i} \frac{s_i^2}{m_i}$$

$$= \frac{50-10}{50(10)(53)^2}(135.85)^2 + \frac{1}{50(10)(53)^2}(4713.2)$$

($\overline{M}$ is estimated by $\overline{m}$)

$$B = 2\sqrt{\hat{V}(\hat{\mu}_r)} = 1.455$$

where, from above summary statistics table,

$$\overline{m} = 53, \quad s_r^2 = \frac{1}{n-1}\sum M_i^2(\overline{y}_i - \hat{\mu}_r)^2 = (135.85)^2$$

$$\sum M_i^2 \frac{M_i - m_i}{M_i}\frac{s_i^2}{m_i} = \sum M_i(M_i - m_i)\frac{s_i^2}{m_i} = 4713.2$$

9.3 Since $M = 2600$ is known, use Equation (9.1) to estimate μ.

$$\hat{\mu} = \frac{N}{M}\frac{\sum M_i\overline{y}_i}{n} = \frac{50}{2600}(497.1) = 9.559$$

The estimated variance of $\hat{\mu}$ is, from Equation (9.2),

$$\hat{V}(\hat{\mu}) = \frac{N-n}{Nn\overline{M}^2}s_b^2 + \frac{1}{Nn\overline{M}^2}\sum M_i^2\frac{M_i - m_i}{M_i}\frac{s_i^2}{m_i}$$

$$= \frac{50-10}{50(10)(52)^2}(125.21)^2 + \frac{1}{50(10)(52)^2}(4713.2)$$

$$B = 2\sqrt{\hat{V}(\hat{\mu})} = 1.367$$

where $\quad \overline{M} = \dfrac{M}{N} = \dfrac{2600}{50} = 52$

$$s_b^2 = \frac{1}{n-1}\sum(M_i\overline{y}_i - \overline{M}\hat{\mu})^2 = (125.21)^2$$

9.6 $N = 12, \ n = 4$

Dept	M_i	m_i	$\overline{y}_i$	s_i^2	$M_i\overline{y}_i$	within
1	21	10	15.5	2.8	325.5	64.68
2	23	10	15.8	3.1	363.4	92.69
3	20	10	17.0	3.5	340.0	70.00
4	20	10	14.9	3.4	298.0	68.00

where *within* is defined as $M_i(M_i - m_i)s_i^2 / m_i$

Summary statistics

	n	mean	stdev
M_i	4	21.00	1.414
m_i	4	10.00	0
$M_i\overline{y}_i$	4	331.725	27.373
within	4	73.842	12.76

$$\hat{\tau} = N\frac{\sum M_i\overline{y}_i}{n} = 12(331.725) = 3980.7$$

$$\hat{V}(\hat{\tau}) = \left(\frac{N-n}{N}\right)\frac{N^2}{n}s_b^2 + \frac{N}{n}\sum M_i^2\left(\frac{M_i - m_i}{M_i}\right)\left(\frac{s_i^2}{m_i}\right)$$

$$= \left(\frac{12-4}{12}\right)\frac{12^2}{4}(27.373)^2 + \frac{12}{4}(295.37)$$

$$B = 2\sqrt{\hat{V}(\hat{\tau})} = 274.73$$

where

$$s_b^2 = \frac{1}{n-1}\sum(M_i\bar{y}_i - \overline{M}\hat{\mu})^2 = (27.373)^2$$

$$\sum M_i(M_i - m_i)\left(\frac{s_i^2}{m_i}\right) = 4(73.842) = 295.37$$

9.7 $N = 7, \ n = 3$

Area	M_i	m_i	$\hat{p}_i$	$M_i\hat{p}_i$	$M_i(\hat{p}_i - \hat{p})$	within
1	46	9	0.111111	5.1111	−0.40862	21.0123
2	67	13	0.153846	10.3077	2.26808	39.2485
3	93	20	0.100000	9.3000	−1.85946	32.1584

where *within* is defined as $M_i(M_i - m_i)\hat{p}_i(1 - \hat{p}_i)/(m_i - 1)$

Summary statistics

	n	mean	stdev
M_i	3	68.6667	23.544
m_i	3	14.0	5.5678
$M_i\hat{p}_i$	3	8.2396	2.7558
$M_i(\hat{p}_i - \hat{p})$	3	0	2.0939
within	3	30.806	9.1830

$$\hat{p} = \frac{\sum M_i\hat{p}_i}{\sum M_i} = \frac{\sum M_i\hat{p}_i/n}{\sum M_i/n} = \frac{8.2396}{68.6667} = .120$$

$$\hat{V}(\hat{p}) = \left(\frac{N-n}{N}\right)\frac{1}{n\overline{M}^2}s_r^2 + \frac{1}{nN\overline{M}^2}\sum M_i^2\left(\frac{M_i - m_i}{M_i}\right)\left(\frac{\hat{p}_i\hat{q}_i}{m_i - 1}\right)$$

$$= \left(\frac{7-3}{7}\right)\frac{1}{3(68.6667)^2}(2.0939)^2 + \frac{1}{3(7)(68.6667)^2}(92.418)$$

$$B = 2\sqrt{\hat{V}(\hat{p})} = .067$$

where

$$s_r^2 = \frac{1}{n-1} \sum M_i^2 (\hat{p}_i - \hat{p})^2 = (2.0939)^2$$

$$\sum M_i^2 \left(\frac{M_i - m_i}{M_i} \right) \left(\frac{\hat{p}_i \hat{q}_i}{m_i - 1} \right) = 30.806 \times 3 = 92.418$$

9.9 $N = 24 \quad n = 6 \quad \overline{M} = 12$

Case	M_i	m_i	$\bar{y}_i$	s_i^2	$M_i \bar{y}_i$	within
1	12	4	7.9	0.15	94.8	3.60
2	12	4	8.0	0.12	96.0	2.88
3	12	4	7.8	0.09	93.6	2.16
4	12	4	7.9	0.11	94.8	2.64
5	12	4	8.1	0.10	97.2	2.40
6	12	4	7.9	0.12	94.8	2.88

where *within* is defined as $M_i(M_i - m_i)s_i^2 / m_i$

Summary statistics

	n	mean	stdev
M_i	6	12	0
m_i	6	4	0
$M_i \bar{y}_i$	6	95.2	1.2394
within	6	2.76	0.498

$$\hat{\mu} = \frac{N}{M} \frac{\sum M_i \bar{y}_i}{n} = \frac{1}{\overline{M}} \frac{\sum M_i \bar{y}_i}{n} = \frac{95.2}{12} = 7.93$$

$$\hat{V}(\hat{\mu}) = \frac{N-n}{N} \frac{1}{n\overline{M}^2} s_b^2 + \frac{1}{nN\overline{M}^2} \sum M_i^2 \left(\frac{M_i - m_i}{M_i} \right) \left(\frac{s_i^2}{m_i} \right)$$

$$= \frac{24-6}{24} \frac{1}{6(12)^2} (1.2394)^2 + \frac{1}{6(24)(12)^2} (16.56)$$

$$B = 2\sqrt{\hat{V}(\hat{\mu})} = .0923$$

where

$$s_b^2 = \frac{1}{n-1} \sum (M_i \bar{y}_i - \overline{M}\hat{\mu})^2 = (1.2394)^2$$

$$\sum M_i(M_i - m_i)\left(\frac{s_i^2}{m_i}\right) = 6(2.76) = 16.56$$

9.11 $N = 20, \ n = 5$

City	M_i	m_i	$\bar{y}_i$	s_i	$M_i\bar{y}_i$	$M_i(\bar{y}_i - \hat{\mu}_r)$	within
1	45	9	102	20	4590	181.225	3600.00
2	36	7	90	16	3240	−287.020	2386.29
3	20	4	76	22	1520	−439.456	1760.00
4	18	4	94	26	1692	−71.510	1638.00
5	28	6	120	12	3360	616.762	1232.00

where *within* is defined as $M_i(M_i - m_i)s_i^2 / m_i$

Summary statistics

	n	mean	stdev
M_i	5	29.4	11.26
m_i	5	5.2	2.121
$M_i\bar{y}_i$	5	288.04	1279
$M_i(\bar{y}_i - \hat{\mu}_r)$	5	0	416.49
within	5	2132.2	924

$$\hat{\mu}_r = \frac{\sum M_i\bar{y}_i}{\sum M_i} = \frac{\sum M_i\bar{y}_i / n}{\sum M_i / n} = \frac{288.04}{29.4} = 97.97$$

$$\hat{V}(\hat{\mu}_r) = \left(\frac{N-n}{N}\right)\frac{1}{n\bar{M}^2}s_r^2 + \frac{1}{nN\bar{M}^2}\sum M_i^2\left(\frac{M_i - m_i}{M_i}\right)\left(\frac{s_i^2}{m_i}\right)$$

$$= \left(\frac{20-5}{20}\right)\frac{1}{5(147/5)^2}(416.49)^2 + \frac{1}{5(20)(147/5)^2}(10616)$$

$$B = 2\sqrt{\hat{V}(\hat{\mu}_r)} = 10.996$$

where

$$s_r^2 = \frac{1}{n-1}\sum M_i^2(\bar{y}_i - \hat{\mu}_r)^2 = (416.49)^2$$

$$\sum M_i(M_i - m_i)\left(\frac{s_i^2}{m_i - 1}\right) = 5(2132.2) = 10616$$

56

9.14 $N = 300, \ n = 4$

Block	M_i	m_i	$\bar{y}_i$	s_i	$M_i\bar{y}_i$	$M_i(\bar{y}_i - \hat{\mu}_r)$	within
1	18	3	1.00000	1.00000	18	0.33962	90
2	14	3	1.00000	1.73205	14	0.26415	154
3	9	3	1.33333	0.57735	12	3.16981	6
4	12	3	0.66667	0.57735	8	−3.77358	12

where *within* is defined as $M_i(M_i - m_i)s_i^2 / m_i$

Summary statistics

	n	mean	stdev
M_i	4	13.25	3.77
m_i	4	4	0.00
$M_i\bar{y}_i$	4	13	4.16
$M_i(\bar{y}_i - \hat{\mu}_r)$	4	0	2.8561
within	4	65.5	70.3

The best estimate of the total number of retired residents in the city is

$$\hat{\tau} = N\frac{\sum M_i\bar{y}_i}{n} = 300(13) = 3900$$

The bound on the error of estimation is

$$B = 2\sqrt{\hat{V}(\hat{\tau}_r)} = 2\sqrt{\hat{V}(N\overline{m}\hat{\mu}_r)} = 2N\overline{m}\sqrt{\hat{V}(\hat{\mu}_r)}$$

$$= 2N\overline{m}\sqrt{\frac{N-n}{Nn\overline{M}^2}s_r^2 + \frac{1}{Nn\overline{M}^2}\sum M_i^2\frac{M_i - m_i}{M_i}\frac{s_i^2}{m_i}}$$

$$= 2(300(13.15)\sqrt{\frac{300-4}{300(4)(13.25)^2}(2.8561)^2 + \frac{262}{300(4)(13.25)^2}}$$

$$= 896.10$$

where $\overline{M}$ is estimated by $\overline{m}$, $\overline{m} = \dfrac{\sum M_i}{n} = 13.25$

$$s_r^2 = \frac{1}{n-1}\sum M_i^2(\bar{y}_i - \hat{\mu}_r)^2 = (2.8561)^2$$

$$\sum M_i(M_i - m_i)\frac{s_i^2}{m_i} = 262$$

9.15 Average number of retired residents per household

$$\hat{\mu}_r = \frac{\sum M_i \bar{y}_i}{\sum M_i} = \frac{\sum M_i \bar{y}_i / n}{\sum M_i / n} = \frac{13}{13.25} = .9811$$

$$\hat{V}(\hat{\mu}_r) = \left(\frac{N-n}{N}\right)\frac{1}{n\overline{M}^2}s_r^2 + \frac{1}{nN\overline{M}^2}\sum M_i^2\left(\frac{M_i - m_i}{M_i}\right)\left(\frac{s_i^2}{m_i}\right)$$

$$= \left(\frac{300-4}{300}\right)\frac{1}{4(13.25)^2}(2.8561)^2 + \frac{1}{4(300)(13.25)^2}(262)$$

$$B = 2\sqrt{\hat{V}(\hat{\mu}_r)} = .2254$$

CHAPTER 10

ESTIMATING THE POPULATION SIZE

10.5 $n = 515, \quad t = 320, \quad s = 91$

$$\hat{N} = \frac{nt}{s} = \frac{515(320)}{91} = 1810.99 \approx 1811$$

$$\hat{V}(\hat{N}) = \frac{t^2 n(n-s)}{s^3} = \frac{(320)^2(515)(515-91)}{(91)^3}$$

$$B = 2\sqrt{\hat{V}(\hat{N})} = 344.51$$

10.7 $n = 750, \quad t = 750, \quad s = 168$

$$\hat{N} = \frac{nt}{s} = \frac{750(750)}{168} = 3348.2 \approx 3349$$

$$\hat{V}(\hat{N}) = \frac{t^2 n(n-s)}{s^3} = \frac{(750)^2(750)(750-168)}{(168)^3}$$

$$B = 2\sqrt{\hat{V}(\hat{N})} = 455.11$$

10.9 $N = 2500$

$$B = 2\sqrt{\hat{V}(\hat{N})} = 356$$

$$\hat{V}(\hat{N}) = \left(\frac{356}{2}\right)^2 = 31684$$

$$\frac{\hat{V}(\hat{N})}{N} = \frac{31684}{2500} = 12.67$$

Using either Figure 10.1 or Table 10.1, let $p_1 = p_2 = .25$

$t = p_1 N = .25(2500) = 625$

$s = p_2 N = .25(2500) = 625$

10.11 $n = 75, \quad t = 100, \quad s = 10$

$$\hat{N} = \frac{nt}{s} = \frac{75(100)}{10} = 750$$

$$\hat{V}(\hat{N}) = \frac{t^2 n(n-s)}{s^3} = \frac{(100)^2(75)(75-10)}{(10)^3}$$

$$B = 2\sqrt{\hat{V}(\hat{N})} = 441.59$$

10.13 $n = 100, \quad t = 120, \quad s = 48$

$$\hat{N} = \frac{nt}{s} = \frac{100(120)}{48} = 250$$

$$\hat{V}(\hat{N}) = \frac{t^2 n(n-s)}{s^3} = \frac{(120)^2(100)(100-48)}{(48)^3}$$

$$B = 2\sqrt{\hat{V}(\hat{N})} = 52.04$$

10.15 $n = 500, \quad y = 500-410$

$$\hat{\lambda} = -\frac{1}{a}\ln\left(\frac{y}{n}\right) = -\frac{1}{100}\ln\left(\frac{500-410}{500}\right) = .0171$$

$$\hat{V}(\hat{\lambda}) = \frac{1}{na^2}\left(e^{\lambda a} - 1\right) = \frac{1}{500(100)^2}\left(e^{(.0171)(100)} - 1\right)$$

$$B = 2\sqrt{\hat{V}(\hat{\lambda})} = .00191$$

10.19 $a = 1$ field, $\quad n = 240$ fileds

$$\sum m_i = 0(11) + 1(37) + 2(64) + 3(65) + 4(37) + 5(24) + 6(12) = 670$$

$$\overline{m} = \frac{\sum m_i}{n} = \frac{670}{240} = 2.792$$

$$\hat{\lambda} = \frac{\overline{m}}{a} = \frac{2.79}{1} = 2.792 \qquad B = 2\sqrt{\frac{\hat{\lambda}}{an}} = 2\sqrt{\frac{2.792}{(1)(240)}} = .216$$

10.21 **(a)** $a = 15$ minutes $= .25$ hours

$n = 10$

$$\sum m_i = 1(0) + 3(1) + 6(2) = 15$$

$$\overline{m} = \frac{\sum m_i}{n} = \frac{15}{10} = 1.5 \qquad \hat{\lambda} = \frac{\overline{m}}{a} = \frac{1.5}{.25} = 6$$

(b) $\hat{V}(\hat{\lambda}) = \frac{\hat{\lambda}}{an} = \frac{6}{(.25)(10)} = 2.4$

CHAPTER 11

SUPPLEMENTAL TOPICS

11.1 $k = 8$

$\bar{y}_1 = 322.6$ $\bar{y}_5 = 404.6$

$\bar{y}_2 = 345.8$ $\bar{y}_6 = 593.8$

$\bar{y}_3 = 493.8$ $\bar{y}_7 = 584.8$

$\bar{y}_4 = 224.0$ $\bar{y}_8 = 287.6$

$$\sum \bar{y}_i = 3257.0 \qquad \sum \bar{y}_i^2 = 1458667.24$$

The estimated mean $\bar{y}$ is, given in Equation (11.2),

$$\bar{y} = \frac{1}{k} \sum \bar{y}_i = \frac{3257.0}{8} = 407.125$$

The estimated variance of $\bar{y}$ is, given in Equation (11.3), then becomes

$$\hat{V}(\bar{y}) = \left(\frac{N-n}{N} \right) \frac{1}{k(k-1)} \left[\sum \bar{y}_i^2 - 2\bar{y} \sum \bar{y}_i + k\bar{y}^2 \right]$$

$$= \left(\frac{545-40}{545} \right) \frac{1}{8(7)} \left[1458667.24 - 2(407.125)(3257.0) + 8(407.125)^2 \right]$$

with $B = 2\sqrt{\hat{V}(\bar{y})} = 93.7$

11.3 $N = 95$, $n = 20$, $n_1 = 16$

$$\sum y_{1j} = 377.78 \qquad \sum y_{1j}^2 = 15209.5336$$

The estimator of the population mean is $\bar{y}_1$, given by Equation (11.5), which yields an estimate of

$$\bar{y}_1 = \frac{1}{n_1} \sum y_{1j} = \frac{377.78}{16} = 23.61$$

The quantity $(N_1 - n_1)/N_1$ must be estimated by $(N-n)/N$, since N_1 is unknown. The estimate of $\bar{y}_1$, given in Equation (11.6), then becomes

$$\hat{V}(\bar{y}_1) = \left(\frac{N-n}{N}\right)\frac{1}{n_1(n_1-1)}\left[\sum y_{1j}^2 - n_1\bar{y}_1^2\right]$$

$$= \left(\frac{95-20}{95}\right)\frac{1}{16(15)}\left[15209.5336 - 16(23.61)^2\right]$$

with

$$B = 2\sqrt{\hat{V}(\bar{y}_1)} = 9.097$$

11.5 When N_1 ($= 83$) is known the estimate of total amount of past-due accounts for the store, is, by Equation (11.7),

$$\hat{\tau}_1 = \frac{N_1}{n_1}\sum y_{1j} = \frac{83}{16}(377.78) = 1959.73$$

The estimated variance of $\hat{\tau}_1$ is, by Equation (11.8)

$$\hat{V}(\hat{\tau}_1) = N_1^2\left(\frac{N_1-n_1}{N_1}\right)\frac{1}{n_1(n_1-1)}\left[\sum y_{1j}^2 - n_1\bar{y}_1^2\right]$$

$$= (83)^2\left(\frac{83-16}{83}\right)\frac{1}{16(15)}\left[15209.5336 - 16(23.61)^2\right]$$

with

$$B = 2\sqrt{\hat{V}(\hat{\tau}_1)} = 763.51$$

11.7 When N_1 is unknown, the estimate of total amount of past-due accounts for the store is, by Equation (11.9),

$$\hat{\tau}_1 = \frac{N}{n}\sum y_{1j} = \frac{493}{30}(235.3) = 3866.763$$

The estimated variance of $\hat{\tau}_1$ is, by Equation (11.10)

$$\hat{V}(\hat{\tau}_1) = N^2\left(\frac{N-n}{N}\right)\frac{1}{n(n-1)}\left[\sum y_{1j}^2 - \frac{1}{n}\left(\sum y_{1j}\right)^2\right]$$

$$= (493)^2\left(\frac{493-30}{493}\right)\frac{1}{30(29)}\left[3136.33 - \frac{1}{30}(235.3)^2\right]$$

with

$$B = 2\sqrt{\hat{V}(\hat{\tau}_1)} = 1163.892$$

11.9 From Equation (11.11) with $\theta = .8$, $n = 200$, and $n_1 = 145$,

$$\hat{p} = \frac{1}{2\theta - 1}\left(\frac{n_1}{n}\right) - \left(\frac{1 - \theta}{2\theta - 1}\right) = \frac{1}{2(.8) - 1}\left(\frac{145}{200}\right) - \frac{1 - .8}{2(.8) - 1} = .875$$

and from Equation (11.12),

$$\hat{V}(\hat{p}) = \frac{1}{(2\theta - 1)^2}\frac{1}{n}\left(\frac{n_1}{n}\right)\left(1 - \frac{n_1}{n}\right) = \frac{1}{[2(.8) - 1]^2}\frac{1}{200}\left(\frac{145}{200}\right)\left(1 - \frac{145}{200}\right)$$

with

$$B = 2\sqrt{\hat{V}(\hat{p})} = .105$$

CHAPTER 12
SUMMARY

12.1 **(a)** For weight, $N = 6000$, $n = 30$

$$\hat{\mu} = \bar{y} = \frac{\sum y_i}{n} = \frac{1929.8}{30} = 64.33$$

$$B = 2\sqrt{\hat{V}(\bar{y})} = 2\sqrt{\frac{s^2}{n}\left(\frac{N-n}{N}\right)} = 2\sqrt{\frac{(1.428)^2}{30}\left(\frac{6000-30}{6000}\right)} = .52$$

The average weight of batteries with a bound on the error of estimation is

$$\hat{\mu} \pm 2\sqrt{\hat{V}(\hat{\mu})} \quad \text{or} \quad 64.33 \pm .52 \quad \text{or} \quad (63.81, 64.85)$$

Since interval does not cover 69, manufacturer's specifications is not met for this shipment.

(b) For plate thickness, $N = 6000$, $n = 30$, $M = 24(6000) = 144000$

$$\overline{M} = \frac{M}{N} = \frac{24(6000)}{6000} = 24, \quad M_i = 24, \text{ for } i = 1, \cdots, 30$$

Battery	M_i	m_i	$\bar{y}_i$	s_i	$M_i\bar{y}_i$	within
1	24	8	109.6	0.74	2630.4	26.285
2	24	16	110.0	1.22	2640.0	17.861
3	24	16	107.0	1.83	2568.0	40.187
4	24	16	111.6	2.55	2678.4	78.030
5	24	17	110.7	1.65	2656.8	26.905
6	24	16	108.7	1.40	2608.8	23.520
7	24	16	111.4	2.63	2673.6	83.003
8	24	13	112.8	2.06	2707.2	86.178
9	24	16	107.8	3.35	2587.2	134.670
10	24	8	109.9	1.25	2637.6	75.000
11	24	16	107.8	3.19	2587.2	122.113
12	24	16	110.2	1.22	2644.8	17.861
13	24	12	112.0	1.81	2688.0	78.626
14	24	12	108.5	1.57	2604.0	59.158
15	24	12	110.4	1.68	2649.6	67.738
16	24	12	111.8	1.64	2683.2	64.550
17	24	12	111.9	1.68	2685.6	67.738
18	24	12	112.5	1.00	2700.0	24.000
19	24	12	109.2	2.44	2620.8	142.886
20	24	12	106.1	2.23	2546.4	119.350

21	24	12	112.0	0.95	2688.0	21.660
22	24	12	112.8	1.75	2707.2	73.500
23	24	12	110.2	2.05	2644.8	100.860
24	24	12	108.0	2.37	2592.0	134.806
25	24	7	112.4	0.79	2697.6	36.376
26	24	12	106.6	2.47	2558.4	146.422
27	24	12	110.5	1.62	2652.0	62.986
28	24	12	113.3	1.23	2719.2	36.310
29	24	12	112.7	1.23	2704.8	36.310
30	24	12	110.6	1.68	2654.4	67.738

where *within* is defined as $M_i(M_i - m_i)s_i^2 / m_i$

Summary statistics

	n	mean	stdev
M_i	30	24	0
m_i	30	12.833	2.561
$M_i \bar{y}_i$	30	2647.2	48.4
within	30	69.09	40.06

The best estimate of μ is $\hat{\mu}$. From Equation (9.1),

$$\hat{\mu} = \frac{N}{M} \frac{\sum M_i \bar{y}_i}{n} = \frac{1}{\overline{M}} \frac{\sum M_i \bar{y}_i}{n} = \frac{2647.2}{24} = 110.3$$

The estimated variance of $\hat{\mu}$ is, from Equation (9.2),

$$\hat{V}(\hat{\mu}) = \left(\frac{N-n}{N}\right) \frac{1}{n\overline{M}^2} s_b^2 + \frac{1}{n N \overline{M}^2} \sum M_i^2 \left(\frac{M_i - m_i}{M_i}\right)\left(\frac{s_i^2}{m_i}\right)$$

$$= \left(\frac{6000 - 30}{30}\right) \frac{1}{30(24)^2} (48.4)^2 + \frac{1}{30(6000)(24)^2} (2072.7)$$

$$B = 2\sqrt{\hat{V}(\hat{\mu})} = .74$$

where

$$s_b^2 = \frac{1}{n-1} \sum (M_i \bar{y}_i - \overline{M}\hat{\mu})^2 = (48.4)^2$$

$$\sum M_i(M_i - m_i)\left(\frac{s_i^2}{m_i}\right) = 30(69.09) = 2072.7$$

The average plate thickness with a bound on the error of estimation is

$$\hat{\mu} \pm 2\sqrt{\hat{V}(\hat{\mu})} \quad \text{or} \quad 110.3 \pm .74 \quad \text{or} \quad (109.56, 111.04)$$

Since interval does not cover 120, manufacturer's specifications is not met for this shipment.

12.4 $N = 180$, $n = 20$ $n_1 = 12$ $n_2 = 8$

	All		F		S	
	Estimate	Actual	Estimate	Actual	Estimate	Actual
$\sum y_i$	5986	6029	33500	3312	2636	27717
$\sum y_i^2$	1993818	2011289	1036188	1017768	957630	993521
s^2	10642.5	10202.5	9177.97	9423.27	12724.0	10108.6

(a)   Total actual board feet (All)

$$\hat{\tau} = N\bar{y} = N\frac{\sum y_i}{n} = 180\frac{6029}{20} = 54261$$

$$\hat{V}(\hat{\tau}) = N^2 \frac{s^2}{n}\left(\frac{N-n}{N}\right) = 180^2\frac{10202.5}{20}\left(\frac{180-20}{180}\right) = 14691600$$

$$B = 2\sqrt{\hat{V}(\hat{\tau})} = 7665.92$$

(b)   Proportion of balsam fir trees in the entire stands (F)

$$\hat{p} = \frac{12}{20} = .6$$

$$\hat{V}(\hat{p}) = \frac{\hat{p}\hat{q}}{n-1}\left(\frac{N-n}{N}\right) = \frac{(.6)(.4)}{19}\left(\frac{180-20}{180}\right) = .011$$

$$B = 2\sqrt{\hat{V}(\hat{p})} = .21$$

(c)   When N_1 (population size for balsam fir) is unknown, the estimate of total actual board feet of balsam fir in the stand is, by Equation (11.9),

$$\hat{\tau}_1 = \frac{N}{n}\sum y_{1j} = \frac{180}{20}(3312) = 29808$$

The estimated variance of $\hat{\tau}_1$ is, by Equation (11.10),

$$\hat{V}(\hat{\tau}_1) = N^2\left(\frac{N-n}{N}\right)\frac{1}{n(n-1)}\left[\sum y_{1j}^2 - \frac{1}{n}\left(\sum y_{1j}\right)^2\right]$$

$$= 180^2\left(\frac{180-20}{180}\right)\frac{1}{20(19)}\left[1017768 - \frac{1}{20}(3312)^2\right] = 35568061$$

with

$$B = 2\sqrt{\hat{V}(\hat{\tau}_1)} = 11927.79$$

(d)   When $N_1\ (= 110,$ population size for balsam fir) is known, the estimate of total actual board feet of balsam fir in the stand is, by Equation (11.7),

$$\hat{\tau}_1 = \frac{N_1}{n_1}\sum y_{1j} = \frac{110}{12}(3312) = 30360$$

The estimated variance of $\hat{\tau}_1$ is, by Equation (11.8),

$$\hat{V}(\hat{\tau}_1) = N_1^2\left(\frac{N_1-n_1}{N_1}\right)\frac{1}{n_1(n_1-1)}\left[\sum y_{1j}^2 - \frac{1}{n_1}\left(\sum y_{1j}\right)^2\right]$$

$$= 110^2\left(\frac{110-12}{110}\right)\frac{1}{12(11)}\left[1017768 - \frac{1}{12}(3312)^2\right] = 8465240$$

with

$$B = 2\sqrt{\hat{V}(\hat{\tau}_1)} = 5819.02$$

12.5   **(a)**   Average age of stone

	All stone	
	Carolinas	Rockies
n_i	830	450
$\bar{y}_i$	43.832	44.155
s_i^2	110.364	107.663

where

$$\bar{y}_1 = \frac{363(42.2)+467(45.1)}{363+467} = \frac{36380.3}{830} = 43.832$$

$$\bar{y}_2 = \frac{259(42.5)+191(46.4)}{259+191} = \frac{1986.99}{450} = 44.155$$

$$s_1^2 = \frac{362(10.9)^2+466(10.2)^2}{829} = \frac{91491.86}{829} = 110.364$$

$$s_2^2 = \frac{258(10.8)^2+190(9.8)^2}{449} = \frac{48340.72}{449} = 107.663$$

N_1 and N_2 are unknown. Assume that they are equal.

Let N' represent both of these terms.

$$\bar{y}_{st} = \frac{1}{N}\sum N_i\bar{y}_i = \frac{1}{2N'}\sum N\bar{y}_i = \frac{1}{2}(\bar{y}_1 + \bar{y}_2) = \frac{1}{2}(43.832 + 44.155) = 43.994$$

$$\hat{V}(\bar{y}_{st}) = \frac{1}{N^2}\sum N_i^2 \frac{s_i^2}{n_i}$$

$$= \frac{1}{(2N')^2}\sum N'^2 \frac{s_i^2}{n_i}$$

$$= \frac{1}{4}\sum \frac{s_i^2}{n_i} = \frac{1}{4}\left(\frac{110.364}{830} + \frac{107.663}{450}\right) = .0931$$

$$B = 2\sqrt{\hat{V}(\bar{y}_{st})} = .62$$

12.5 **(b)** Average calcium concentration in Carolinas

$$\bar{y} = \frac{363(11.0) + 467(11.3)}{830} = 11.169$$

$$s^2 = \frac{362(15.1)^2 + 466(16.6)^2}{829} = 254.464$$

$$\hat{V}(\bar{y}) = \frac{s^2}{n} = \frac{254.46}{830} = .307$$

$$B = 2\sqrt{\hat{V}(\bar{y})} = 1.107$$

(c) Average calcium concentration in Rockies

$$\bar{y} = \frac{259(42.4) + 191(40.1)}{450} = 41.424$$

$$s^2 = \frac{258(31.8)^2 + 190(28.4)^2}{449} = 922.375$$

$$\hat{V}(\bar{y}) = \frac{s^2}{n} = \frac{922.375}{450} = 2.05$$

$$B = 2\sqrt{\hat{V}(\bar{y})} = 2.86$$

Average calcium concentration in drinking water differ between two areas because confidence intervals do not overlap each other.

(d) proportion of smokers among new stones

	Carolinas	Rockies
$\hat{p}_i$	.73	.57
n_i	830	450

$$\hat{p}_{st} = \frac{1}{N}\sum N_i\hat{p}_i = \frac{1}{2N'}\sum N'\hat{p}_i = \frac{1}{2}\left(\hat{p}_1 + \hat{p}_2\right) = \frac{1}{2}(.73+.57) = .65$$

$$\hat{V}(\hat{p}_{st}) = \frac{1}{N^2}\sum N_i^2 \frac{\hat{p}_i\hat{q}_i}{(n_i-1)} = \frac{1}{(2N')^2}\sum (N')^2 \frac{\hat{p}_i\hat{q}_i}{(n_i-1)}$$

$$= \frac{1}{4}\sum \frac{\hat{p}_i\hat{q}_i}{(n_i-1)} = \frac{1}{4}\left(\frac{(.73)(.27)}{830} + \frac{(.57)(.43)}{450}\right)$$

$$B = 2\sqrt{\hat{V}(\hat{p}_{st})} = .028$$

12.8 $n = 12,\quad x = \text{LC50 Static},\quad y = \text{LC50 Flow-through}$

$$\sum x_i = 137.3 \qquad \sum x_i^2 = 8215.97$$

$$\sum y_i = 126.3 \qquad \sum y_i^2 = 7172.97 \qquad \sum x_i y_i = 2221.48$$

$$r = \frac{\sum y_i}{\sum x_i} = \frac{126.3}{137.3} = .9198$$

$$\hat{V}(r) \approx \left(\frac{1}{n}\right)\left(\frac{1}{\bar{x}^2}\right)\frac{1}{n-1}\left[\sum y_i^2 - 2r\sum x_i y_i + r^2 \sum x_i^2\right]$$

$$= \left(\frac{1}{12}\right)\frac{1}{(137.3/12)^2}\frac{1}{11}[7172.97 - 2(.9198)(2221.48) + (.9198)^2(8515.97)]$$

$$= .5949$$

$$B = 2\sqrt{\hat{V}(r)} = 1.54$$

12.15   **(a)**

	Operation	Down
1	y_1	x_1
2	y_2	x_2
.	.	.
.	.	.
n	y_n	x_n

use ratio estimate for the proportion of time that the machine is in operation.

$$r = \frac{\sum y_i}{\sum x_i}$$

$$\hat{V}(r) = \left(\frac{1}{n}\right)\left(\frac{1}{\mu_x^2}\right)\frac{1}{n-1}\left[\sum y_i^2 - 2r\sum x_i y_i + r^2 \sum x_i^2\right]$$

$$B = 2\sqrt{\hat{V}(r)}$$

(b) For the total time number of hours of machine operation,

$$r = \frac{\sum y_i}{\sum x_i} \qquad \hat{\tau}_y = r\tau_x = r(40)$$

$$\hat{V}(\hat{\tau}_y) = N^2\left(\frac{N-n}{nN}\right)\frac{1}{n-1}\left[\sum y_i^2 - 2r\sum x_i y_i + r^2 \sum x_i^2\right]$$

$$B = 2\sqrt{\hat{V}(\hat{\tau}_y)}$$

12.17 x = weight $\quad y$ = height

	Boys	Girls
$\sum x_i$	361.0	361.5
$\sum x_i^2$	11145.26	11401.27
$\sum y_i$	1683.4	1678.4
$\sum y_i^2$	211531.6	210440.1
$\sum x_i y_i$	47387.22	47569.06
n_i	14	14

(a) Ratio of height to weight for boys

$$r = \frac{\sum y_i}{\sum x_i} = \frac{1683.4}{361} = 4.663$$

$$\hat{V}(r) = \left(\frac{1}{n}\right)\left(\frac{1}{\mu_x^2}\right)\frac{1}{n-1}\left[\sum y_i^2 - 2r\sum x_i y_i + r^2 \sum x_i^2\right]$$

$$= \left(\frac{1}{14}\right)\frac{1}{(361/14)^2}\frac{1}{13}[211531.6 - 2(4.663)(47387.22) + (4.663)^2(11145.26)]$$

$$B = 2\sqrt{\hat{V}(r)} = .628$$

(b) Ratio of height to weight for girls

$$r = \frac{\sum y_i}{\sum x_i} = \frac{1678.4}{361.5} = 4.643$$

$$\hat{V}(r) = \left(\frac{1}{n}\right)\left(\frac{1}{\mu_x^2}\right)\frac{1}{n-1}\left[\sum y_i^2 - 2r\sum x_i y_i + r^2 \sum x_i^2\right]$$

$$= \left(\frac{1}{14}\right)\frac{1}{\left(\frac{361.5}{14}\right)^2}\frac{1}{13}[210440.1 - 2(4.643)(47569.06) + (4.643)^2(11401.27)]$$

$$B = 2\sqrt{\hat{V}(r)} = .691$$